# Smart Metering Handbook

# SMART METERING HANDBOOK

FABIO TOLEDO

PennWell

PennWell Corporation
1421 South Sheridan Road
Tulsa, Oklahoma 74112-6600 USA

800.752.9764
+1.918.831.9421
sales@pennwell.com
www.Fire EngineeringBooks.com
www.pennwellbooks.com
www.pennwell.com

Marketing Manager: Amanda Alvarez
National Account Executive: Barbara McGee

Director: Mary McGee
Managing Editor: Stephen Hill
Production Manager: Sheila Brock
Production Editor: Tony Quinn
Book Designer: Susan E. Ormston
Cover Designer: Charles Thomas

This is the English language version. Lavoisier Publishing has translation rights for all other languages.

Contact: Lavoisier Bookseller, 14 rue de Provigny, 94236 Cachan Cedex, France
Tél.: 33 (0)1.47.40.67.00
Fax: 33 (0)1.47.40.67.02

Library of Congress Cataloging-in-Publication Data

Toledo, Fabio.
Smart metering handbook / Fabio Toledo.
pages cm
Includes bibliographical references and index.
ISBN 978-1-59370-298-4
1. Electric meters--Handbooks, manuals, etc. 2. Smart power grids--Handbooks, manuals, etc. I. Title.
TK351.T65 2013
621.37'3--dc23

2013007378



Printed in the United States of America

1 2 3 4 5 17 16 15 14 13

I dedicate this book to my beloved parents, Iara and Henrique, to my loving wife, Erica, and to everyone who gave me valuable advice, understanding, patience, support, and who sacrificed their leisure time for the benefit of this work. I also dedicate it to all the other members of my family and friends and colleagues who directly or indirectly contributed to the success of this book.

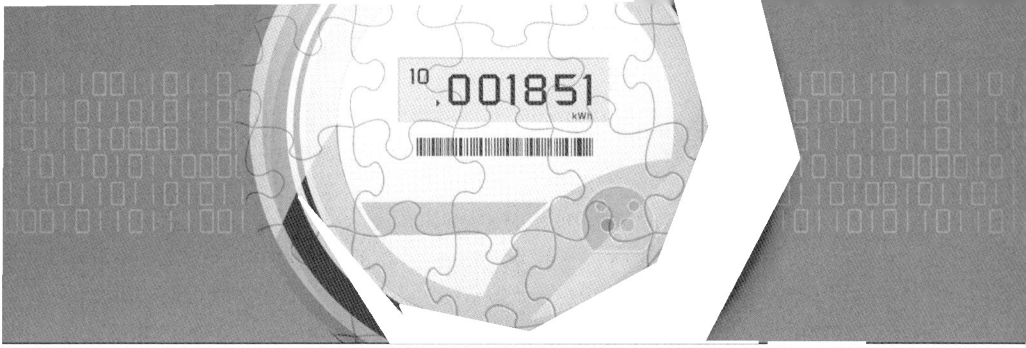


# Contents

# Foreword

In Brazil, there are only a few active companies that are more than a hundred years old. One of these is Light—the electric power company in Rio de Janeiro. Light began as a private Canadian company and subsequently became a Brazilian government–controlled entity. In 1996, Light was privatized, and control was acquired by the French utility company EDF. In 2006, Light became a private Brazilian company with shares traded on the stock market.

Fabio Toledo was 14 years old when he began working at Light. At that time, the company was still state-owned. He went through all the stages of the long process of professional development, starting as an apprentice and eventually becoming the manager of the Metering Department (during the French administration), responsible for new technologies and revenue protection for large accounts. He learned by practicing and studying—practicing and studying a lot!

Energy theft has been one of the faces of the decadence of Rio de Janeiro since the capital moved from Rio to Brasilia in the early 1960s. Fighting this social plague then as well as now is a difficult and, in some circumstances, dangerous task. In the early 2000s in Rio, the government found itself upable to govern some of the poor areas of the city, known as Favelas, and criminal gangs took control of those areas. Some large energy consumers were also stealing high volumes of energy from the distribution network, to the extent of hundreds of thousands of U.S. dollars per month. Light employees faced risky situations fighting energy theft. Today the risk is lower, but still present.

More than an economic challenge to receive fair payment for goods and services delivered, the challenge is also a social one. In the Favelas, where the power utilities have the most problems with theft, there is also what economists call the "tragedy of the commons." There, citizens feel the access to the product is a social right and one that can be freely accessed without concerns about costs or consequences to neighbors or society. When local inhabitants seek only their individual interests, performing unauthorized or illegal activities within the electricity distribution network, it contributes to a collective tragedy. In this case, energy services in areas affected by this social behavior are much worse than those in regularly serviced areas. In the Favelas, where criminal gangs are dominant, there are frequent and long outages, followed by voltage fluctuations that damage electrical appliances.

The electric company found itself engaged in a "guerrilla war" to reverse this situation. Fabio was continually followed by bodyguards to protect him from the wrath of those who felt harmed by the suppression of energy theft. This protection was not enough to prevent an attack on Fabio's car. Several shots hit the vehicle, but fortunately not him. The French executives of Light were understandably concerned about this event, and considering Fabio's expertise acquired throughout his career at Light, they decided to send him to Paris to work in the Research and Development Company of EDF.

At EDF's research and development division, Fabio had an opportunity to improve his skills and knowledge of the concepts of smart metering and learned about a technological novelty—the smart grid, which appeared to attend to the specific needs of electric power companies in Europe, finding solutions to accommodate new forms of renewable energy generation, mainly wind and solar. These forms of energy generation were either concentrated in large production plants or scattered throughout the territory within small consumer premises.

In this new reality, consumers also become producers, but are still dependent on the electricity network, as both wind and solar sources are intermittent. To avoid affecting the reliability of the system, for each new kilowatt from these new plants—which may or may not be producing depending on weather—another kilowatt from installed conventional sources must function as backup. From both an economic or operational point of view, this is not an ideal situation.

In response to this problem, several solutions were developed to make the electricity networks dynamic—networks with changing topologies in real time depending on inputs from energy generation and energy consumption patterns. New concepts based on new techniques in the fields of electrical engineering, communication, and optimization led to the automation of substations, allowing consumers to know in real time the price of electricity, which can vary almost instantaneously according to the marginal cost of producing it. These include bidirectional metering to account for the power flow in consumer-producer units and remote command-and-control capabilities to allow remote connections and disconnections.

Fabio's expertise in the themes and challenges encountered by EDF R&D in France led to assignments in other countries, including South Africa. There, he was involved in the development of a technological breakthrough of great interest for countries with large amounts of poverty, as in the case of Brazil: prepayment meters.

In 2009 Fabio returned to Brazil and to his former company as advisor to the distribution board. I had the opportunity to talk with him and instantly realized the immense potential of this young professional to help solve the same problem that had forced him to move away some years before, but attacking on the other flank: technological advancement.

As president of Light, I promoted Fabio to chief technology officer and tasked him to create a comprehensive smart grid program within a budget of R$36 million from Light's R&D. The first and main step of this program was to develop, within two years, a new metering system smart enough to be competitive with those in Europe, one in which the management capabilities of energy consumption is of the utmost importance. But, above all, this could become an essential tool for fighting energy theft in Brazil and in other developing countries, as it is economically viable, tamper-proof, and capable of remote command and control.

I had a responsibility to maximize the creative potential of Fabio and ensure that he would have the freedom to reach the goals that we defined together. The results we've achieved thus far allow me to be very optimistic.

Jerson Kelman
President of Light (March 2010 to July 2012)
General Manager of ANEEL—Brazilian Regulatory Agency for Electric Energy, January 2005–January 2009
General Manager of ANA—Brazilian Regulatory Agency for Water, December 2000–December 2004

# Preface

Energy suppliers and utilities throughout the world are making decisions to deploy smart metering systems for their customers on a grand scale. The drivers are global incentives for reduced energy consumption and carbon emissions, opening of energy markets, strong pressure of regulators in several areas of energy management, and growing customer demand for new metering system services.

Energy suppliers and utilities have an even stronger perception of their metering systems as necessary and strategic. Opportunities with these systems to retain existing customers and gain new ones are increasingly evident.

A smart metering system project requires multidisciplinary teams (marketing, regulation, metering, R&D, finance, etc.) who are well trained and understand both local challenges and those faced by smart metering deployments throughout the world. This book aims to fill a gap in the available global literature on this subject and to meet the high expectations of these professionals.

For this reason, the book uses language that is meaningful to all who might develop or work on these systems. Nevertheless, the reader should have a basic understanding of the energy market and the metering services of an energy utility, as well as IT, telecommunications, and measurement systems.

As such, the book is intended for a wide audience of professionals who may interact with the subject of smart metering and its opportunities—in universities, energy distributors, energy suppliers, research institutes, standard-setting bodies, laboratories, regulators, metering system manufacturers, suppliers of metering system equipment and components, consulting firms, information system integrators, and other sectors. My goal is for this book to become recommended reading for:

- Technical and R&D managers
- Project managers
- Consultants
- Executives
- Researchers
- Engineers
- Teachers

- Technicians
- Students

I must point out that although this book involves, in principle, both gas and electricity metering systems, its focus is mainly on electricity metering, given its important role in smart energy metering systems. Hence, although other energy (e.g., heat) or nonenergy (e.g., water) metering systems could use some of the concepts in this book, they are beyond its intended scope.

Figures and related information in this book are for illustrative purposes only, not for any comparison between different solutions and companies. The same applies to any other information about companies and deployments mentioned in this book. I do not intend to promote any company or solution. All the companies mentioned are somehow involved with smart metering projects. If for any reason their smart metering solutions (pictures, information, etc.) are not published in this book, this does not mean they do not exist or that a particular company is less qualified than another. The same applies for any company not mentioned here. Because of the large number of companies involved with smart metering projects in the world, it is almost impossible to mention all of them.

It is also important to highlight that although the term *smart metering* is widely employed internationally to refer to the central subject of this book, other terms may be used for the same purpose by different participants of the energy market throughout the world. Examples are the terms *communicating meters* and *advanced metering*.

Also, I cannot assert that this work is complete, as international markets and technology evolve quickly. I could add many other subjects. However, in my opinion, it meets what I intended it to achieve. I thank all my colleagues for their interest and assistance, prior to publication.

Finally, it was a pleasure to write this book and I hope you enjoy it. Any constructive criticisms and suggestions will be gratefully accepted.

# Acknowledgments

Writing this book required a great deal of determination, dedication, and energy, and its publication would have not been possible without the direct or indirect contributions and/or encouragement of a large number of individuals and companies.

I sincerely thank EDF R&D for the opportunity, contributions, and support, especially the members of the Edition Committee and to Sylvie Anglade, Eric Schultz, Stanislas Iweins, and Sebastien Ruiz.

I equally sincerely thank EDF, Light, ERDF, EDF Energy, and EnBW for their direct or indirect support, contributions, and help throughout my career, especially to Steve Hayfield, Ashley Pocock, Francois Blanc, Gilbert Combe, Marco Donatelli, Maria Bernadette Alves, Martial Monfort, Nickolas Slocombe, and Robert Gibbs.

I also sincerely thank all the advisors, reviewers, translators, and editors for their contributions to this book, in particular Theophanis Calliacoudas, Flávia Gouvêa, Ana Paula Jansen, and Alan Knight-Scott, and all the companies that contributed with figures and information.

I also wish to acknowledge the support of all the other contributors who directly or indirectly contributed to the success of this work.

Finally, I would like to thank to Dr. Jerson Kelman for honoring me by writing the preface of this book.

# Introduction

## Smart Metering System Projects: International View

Analysis of the global energy market, with its thousands of distributors and suppliers and its millions of customers, suggests that it is difficult to ensure proper management of all customers without a smart metering system, whether it is a market open to competition (completely or partially) or not.

Until recently, almost all participants throughout the world managed metering systems manually. This results in significant losses, particularly those associated with incorrect customer billing, inefficient operational management of metering equipment, and underutilization of the functions of these systems and associated opportunities.

## Advantages versus Constraints

Some of the problems that hinder the widespread deployment of these systems by distributors and suppliers are related to the difficulty of making these deployments profitable and selecting, among the available technologies, the most appropriate technology for specific features of the market.

It would be tempting to imagine that the smart metering systems and services that these systems can offer to customers are treated as one of the top priorities of all of the world's energy participants, but unfortunately, this is not the reality.

Some energy participants believe their current manual metering systems are sufficient to ensure good management of the energy supplied to their customers, even if they generate a high level of financial loss. These losses vary depending on, among other things, the type of energy supplied, location and type of networks (overhead or underground lines), environment, management organization, and customers. Losses are generally caused by a lack of control, monitoring, and responsiveness.

Smart energy management, through effective deployment of smart metering systems, can solve many of these problems. It is also important to emphasize that the simple, widespread deployment of these systems will not by itself solve all of these problems. Smart energy management and generated data must be used properly for the benefit of customers, regulators, and all of the market's participants.

Certain suppliers also have a reluctant approach that often restricts energy to a single public utility that is unable to generate associated services. However, this view tends to fade with the emergence of the new needs of energy market participants. These needs are nurturing a new way of viewing the subject and are driving the global industry to produce new, adapted technologies. The main phenomena currently observed and producing new needs are described in the following paragraphs.

The opening of markets to competition, full or partial, has generated the need for new services to offer to consumers. These offers allow energy suppliers to retain their customers, gain new customers, and compete with new market entrants. Global requirements to reduce energy consumption and carbon emissions result in the need for effective management of customer demand and changing consumption patterns, as well as the requirements of regulators (regarding the quality of energy supply and the reduction of transport prices). The risks include poor earnings for certain participants due to operational costs, losses, and uncontrolled debt. Other considerations include the emergence of new offerings by indirect competitors, even in closed markets, such as offers of decentralized energy production or services for energy management for end consumers; the emergence of new consumption practices, with different profiles and needs, such as rechargeable electric and hybrid vehicles; aid, sometimes financial, from governments so that network participants can deploy smart metering systems to benefit their customers; and the recent energy supply problems and the resulting demand control measures that have generated a need to bring the consumption curve (load curve) of customers into line with the profile of production and the various associated production costs.

These needs have been the drivers for suppliers and distributors to choose mass deployment (or investigations into the possibility of doing so) of smart metering systems for millions of residential and business customers. These energy participants have realized the strategic importance of metering systems, which is reflected in the use of various technologies and technical architectures for implementation worldwide.

# Methodology and Structure of the Book

While this book deals with dual-energy systems (gas and electricity), its focus is on electricity platforms because of their essential functions in this new environment of smart metering. Various topics relating to this concept are discussed in the chapters that follow.

Chapter 1 provides a technical overview of concepts, metering technologies, and their evolution over time. The main processes are also presented, such as installation, nontechnical losses, payment methods, and information technology (IT) infrastructure.

Once this basic information is acquired, chapter 2 addresses smart metering systems, their technologies, and the technical architecture options. It examines the components of these systems in detail as well as the challenges to overcome, such as interoperability, safety, and data management.

Chapter 3 develops an international view of these platforms and the reasons why they have been implemented on a large scale throughout the world. Their innovative products and services are presented in addition to the main challenges and constraints facing the deployment of these platforms.

Chapter 4 provides the necessary information about implementing technical solutions and mitigating the risks associated with these deployments. A complete end-to-end view of this subject is given in this chapter.

Finally, chapter 5 addresses compatibility with future platforms to be developed. It also discusses trends in these solutions in terms of integration and technological change.

# Summary

The motivations and profitability of the deployment of smart metering systems vary from one market to another, but this global analysis suggests that these systems are already an international reality. More and more distributors and suppliers around the world are adopting smart metering systems to manage their millions of customers more effectively. The new requirements of these markets demonstrate more strongly than ever how strategic a metering system becomes for participants.

Customers and various market participants can also enjoy the benefits associated with the deployment of these systems and related services if these projects are properly deployed and if certain constraints are overcome.

Total customer satisfaction should be the main objective of these projects. As such, it is essential to understand the customers' real needs and expectations. It is also essential that the interface with the customers be simple and adapted to the environment and the customers' real needs to ensure their support for the project.

The functionalities and new services must also be made available to customers during the various phases of the project so that they can enjoy all of the benefits of the project.

The deployment of these systems involves a change in organizational culture, and the success of the projects depends on how this change is executed. The company must therefore provide the necessary resources so that a training program and communications are established in accordance with local laws, the needs of the companies, and the personnel involved so that they are motivated and take part in the cultural and technological change.

All of these topics are addressed in this book in order to help the reader understand the environment of smart metering system platforms.

# Energy Metering Systems

Energy has always been present in our daily lives and naturally available for our use. Early in human history, natural forms of energy, such as lightning from the sky, geothermal sources, and natural emissions of different kinds of gases awoke our curiosity about these phenomena. We soon realized the potential and opportunities associated with these energy sources. Ever since, we have been trying to understand the different effects and behaviors of energy when isolated or in contact with other materials.

Over the past hundred years, we have refined the methods for safely capturing and profiting from different kinds of fuels. Electricity, gas, and other types of fuels have become an everyday part of our lives, and we have been become more and more dependent on them. Many companies were created in order to explore energy and make it available for our use; however, some unintended consequences have also arisen. For example, investments should produce benefits for both business owners and society. Energy technology must be long-lasting and reliably available to customers. We should also not forget that fossil fuels, which are now the primary energy sources most commonly used in some countries, are not inexhaustible and have a significant impact on the environment. This reinforces the need for better control of energy consumption. In this context (economic, technological, environmental, and social), the need to measure energy has become a necessity.

After countless tests and much research effort, standard units were established, measurement equipment was developed, and different market rules were created in order to establish all the necessary parameters for monitoring energy consumption. Meters became a core function in this system. They are the main interface between customers and energy utilities that accurately account for the amount of fuel used by customers. Meters must also control the use of energy along different stages of its production-to-consumption chain. For example, in the electricity domain, meters are installed at strategic points within the electricity network to provide feedback about energy consumption during generation, transmission, and distribution.

It is clear how important meters are for both energy utilities and their customers. Because of its importance and complexity, this subject deserves deep study to understand how it works and the large number of issues associated with metering systems. This chapter focuses on some aspects of electricity and the gas meters, although the main emphasis is on the electricity meters due to their core role in smart metering systems. Information about their main features and services is presented, with an emphasis on how important these systems are for the energy market and on the potential opportunities. This chapter provides the essential background needed to better understand the scope of smart metering systems in subsequent chapters of this book.

# Electricity Measurements Overview

To help the reader with basic understanding about the subjects to be discussed within this chapter, we begin our study by recalling some high-level electricity principles. While an in-depth treatment of these principles is beyond the scope of this book, if you would like to investigate them in greater detail, see the bibliography at the end of the book.

## Electric power

Electric power is the basis for electricity measurement because electric energy can be defined as the quantity of power transferred in a circuit during a specific time. In practical terms, one of the features of electric equipment is its specific value of power. The consumption of electric equipment varies proportionally to its power and the amount of time it is used. Basically, a piece of equipment using higher power than another piece of equipment consumes more energy over the same period. A piece of equipment with power equal to another piece requires more electricity if energized for a longer period of time. For efficient use of equipment, two points should be taken into account: power and time.

In energy-efficiency terms, taking long showers using an electric water heater should concern you more than using an electric shaver over the same period. This is because the electric water heater uses several kilowatts (thousands of watts) of power, while a shaver uses just a few watts. In this example, the water heater consumes many thousand times more energy over the same period.

There are two basic kinds of circuits in the electricity domain: direct current (DC) and alternating current (AC). As a simple explanation, DC may be employed for batteries used to supply a portable radios, and AC is generally used within the home for supplying appliances connected to electric sockets.

AC electric power is classified as active, reactive, and apparent. The power triangle (fig. 1–1) expresses the relationship among them.

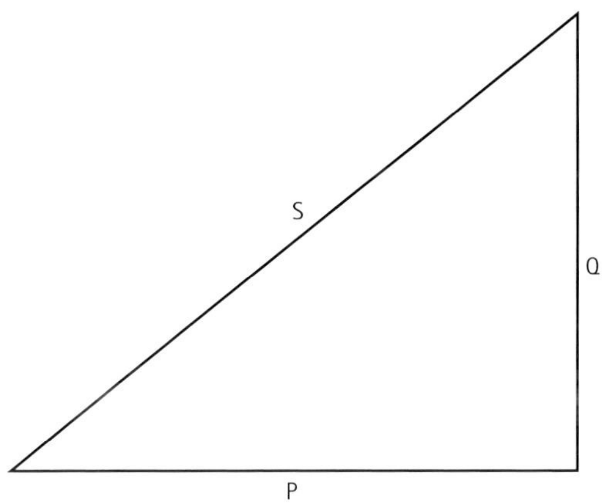


Fig. 1–1. Power triangle (S = apparent power; Q = reactive power; P = active power)

For practical purposes, for a single-phase AC circuit, it is defined that:

- Active power is the amount of power being used or the caloric value dissipated in a circuit. It is the product of voltage, current, and cos $\varphi$ and is expressed in watts (W). In a fully resistive circuit, formed by resistive load only (e.g., a circuit that individually supplies an appliance), cos $\varphi$ is equal to 1. It is also known as average or real power. Its basic formula is

$$P = V \times I \times \cos \varphi$$

where:

$P$ = electric power (watts)
$V$ = voltage (volts)
$I$ = current (amperes)
$\varphi$ = phase angle between $V$ and $I$

- In reactive circuits composed of inductive or capacitive loads, such as appliances that contain coils or capacitors, reactive power is generated. This is the power exchanged between the generation and the load without being consumed. In practical terms, this extra power is not used to provide energy to appliances but is stored in the magnetic circuits (transformers, electric motors) or electric fields (capacitors) and is returned to the power source (the grid) at each alternation (change of polarity) of the current. It is therefore power that permanently circulates between the source (the grid) and load (equipment supplied) without ever being consumed. Of course, this power is undesirable because it is only an overcharge on the grid without ever constituting useful power. Compensation measures are often necessary to minimize or even cancel out that power.

  Reactive power is also the product of voltage, current, and sin φ. It is expressed in volt-amperes reactive (VAr). Reactive power is also known as imaginary power. Its basic formula is

$$Q = V \times I \times \sin \varphi$$

where:

$Q$ = reactive power (VAr)
$V$ = voltage (volts)
$I$ = current (amperes)
$\varphi$ = phase angle between $V$ and $I$

- Apparent or complex power is the vector addition between active and reactive powers. This is the product of voltage and current. It is expressed in volt-amperes (VA), and its basic formula is

$$S = V \times I$$

where:

$S$ = apparent power (VA)
$V$ = voltage (volts)
$I$ = current (amperes)

According to the formulas presented above, apparent power is related to active and reactive power by the following formula (orthogonally between *P* and *Q*):

$$S^2 = P^2 + Q^2$$

Note: All formulas presented above, based on trigonometric properties, are valid for voltage and current that follow a sinusoidal shape, that is, in the absence of harmonics.

Given that the basic units defined by the International System of Units (SI) are sometimes unsuitable for expressing large values, we use kilo-units (1,000 units), mega-units (1,000 kilo-units), and higher orders successively. These are generally used for electricity bills.

The wattmeter (fig. 1–2) is the device used to measure active electric power. Based on the previous principles, in general, we need to have two circuits in order to record active power: current and voltage. The mutual interaction between them (cos $\varphi$) produces the measurement in an active power unit. For energy utilities, they form the basis for the measurement of electric demand.

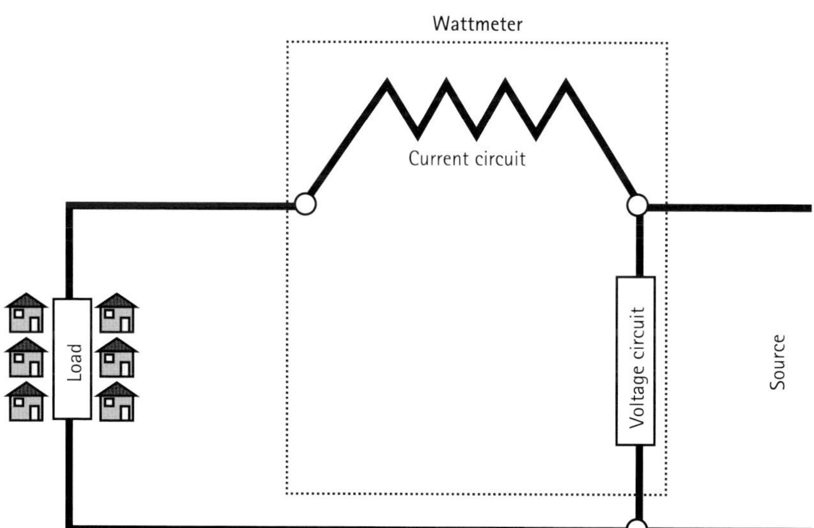


Fig. 1–2. Wattmeter diagram

## Electric energy

As mentioned, electric energy is the power available during a period of time. As such, it can be active, reactive, and apparent. Then,

$$W = (P \text{ or } Q \text{ or } S) \times t$$

Where:

$W$ = energy (kilowatt hour [kWh] or (kilovolt-ampere reactive hour) [kVArh] or kilovolt-ampere hour [kVAh])

$P$ or $Q$ or $S$ = power (kW or kVAr or kVA)

$t$ = time (hours)

For practical illustrative purposes, assuming a fully resistive electric appliance, with load of 1 kW and uninterrupted energized load over a period of 2 hours, the energy consumed is

$$W = P \times t = 1 \times 2 = 2 \text{ kWh}$$

For residential electricity billing purposes, in general, utilities all over the world use the unit kWh. In some countries, specific types of customers may have billing based on both kWh and kVArh.

The equipment used for measuring electric energy is the electricity meter, discussed next.

## Electric demand

In terms of energy metering beyond electric energy measurement, electric demand is another aspect to be recorded. Demand is the power measured within a predefined time interval. The majority of countries have adopted the standard of quarter-hour or half-hour intervals for these registers, and they are generally expressed in kilo-units.

Maximum demand is recorded within an established interval, for example, one month. Demand is measured by utilities so they can accurately build and maintain network infrastructures. Assuming two customers have consumed the same amount of energy in a given month, the first customer may have used a maximum demand of 50 kW within this period, while the second used just 6 kW. The first customer probably has used higher-power equipment in shorter periods of time compared to the second customer. In other words, the first customer had peaks of consumption higher than the second, which had a more regular consumption (smoother consumption profile). Even if both consumed the same amount, the investment in the network for providing energy to the first would generally be more important than to the second in terms of cables, metering systems, and other equipment. This illustrates why energy utilities need to measure maximum demand and try to associate the cost of infrastructure to its specific users.

# Energy Metering Systems

## The history of gas and electricity meters

The gas industry was established early in the 19th century as a competitor to oil. Initially, it was deployed to light the streets of urban areas for short periods. From that simple beginning, gas technology has evolved such that efficient storage makes it possible to use gas for such various purposes as cooking, boiling water, and heating.

Samuel Clegg, chief engineer at the Gas Light and Coke Company in London, developed the first efficient gas meter in 1817. Since then much effort has been put into reducing the size of the meter, containing the cost, and addressing other associated issues.

In 1843, Thomas Glover invented the original diaphragm meter, and in 1870, T. S. Lacey invented the first gas prepayment meter (fig. 1–3).

Around 1870, patents for the first electricity meters were published around the world. At that time, the main competitor of electricity was gas. Just as now, it was important to accurately measure electricity consumption. Although it is hard to determine who invented the first robust electricity meter in the world, Thomas Alva Edison stands out among inventors. He was the first to register an ampere-hour meter for DC applications in 1881, known as a chemical meter. This device was deployed in New York around 1882.

Meylan and Rechniewski built a watt-hour meter for measuring DC circuits in 1886. In consecutive years many other innovative meters were developed and issues resolved. An example of these improvements was the reduction of the amount of energy required by meters for measuring purposes (self-consumption). We could say that the older meters were more energy hungry.

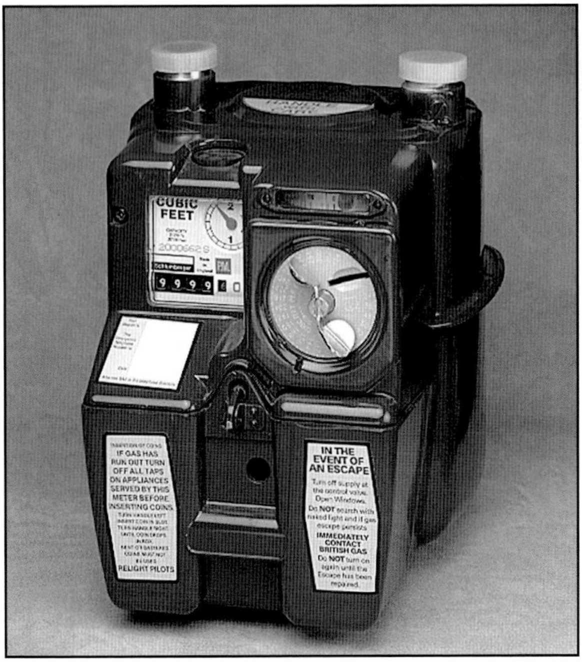


Fig. 1–3. Example of prepayment coin-operated gas meter. (Photo courtesy of Actaris.)

Two different devices of measuring electricity were available: ampere-hour meters and watt-hour meters. After innumerable discussions, the second was established as standard for electricity measurement devices.

Another decision was also required because two different technologies for energy generation were also available: DC and AC. Which one should be chosen to become the standard around the world for residential and other applications? As you know, AC was finally adopted for residential and industrial implementations and DC only for specific applications.

In 1889, the first meters for AC measurements were developed. Elihu Thompson built a watt-hour meter capable of concurrently measuring DC and AC.

Olivier B. Shallenberger created a watt-hour meter as an upgrade to his original ampere-hour version. In 1896, he invented the first polyphase meter.

Over time, other inventions were developed (fig. 1–4) such as the first methods for reactive energy measurement and prepayment meters. Meters able to register electric power demand in kW, kVA, and kVAr were also important innovations, meant to meet the industry needs for multiregister applications.

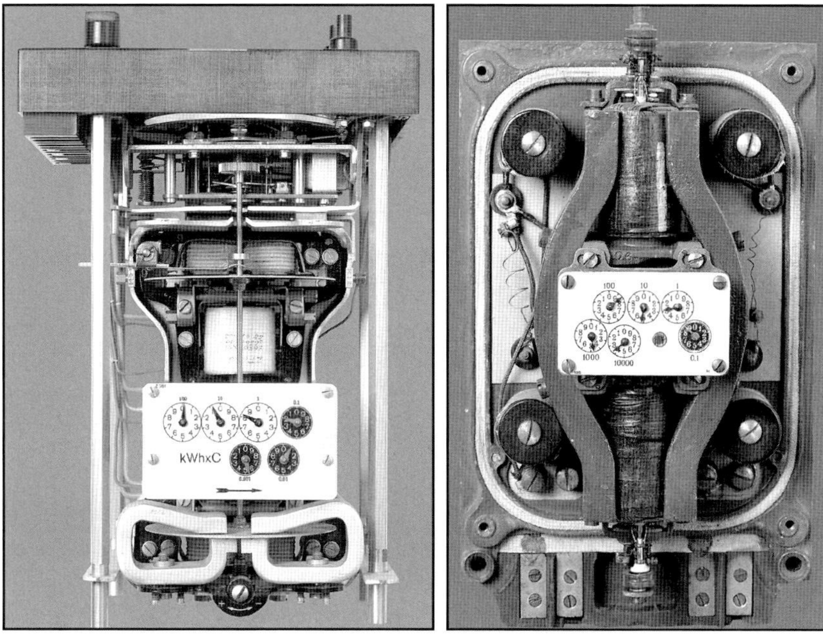


Fig. 1–4. Examples of early electricity meters. (Photo courtesy Landis+Gyr.)

Several meter manufacturers have emerged whose technology has evolved to meet industry requirements. Equipment using new technology is now available, such as electronic energy meters and auxiliary measurement equipment, including instrument transformers.

New measurement devices have been appearing in the market, such as orifice and ultrasonic meters for gas and shunt meters for electricity. Compensation methods and standards have been established, and measurement and regulation guidelines devised.

Nowadays, generally two kinds of meter technologies are available in the market: induction electromechanical and static electronic. Both types of meters are considered integrator devices, measuring the power and energy consumed during a period of time.

In the past, meters were used mainly for billing and energy network monitoring purposes. Now, different customer needs such as energy management and cogeneration call for their use as submeters for various applications in premises, such as for cogeneration monitoring, real-time energy management, load shifting, circuit supervision, demand forecasting, air conditioning and heating equipment monitoring, and individual in-home socket measurement (smart plugs). They have also been used by utility providers for network management as substation measurements, feeder load supervision, demand-side management, quality analysis, energy balance for transformers, special rates, demand forecast, feedbacks for monitoring centers, load curtailment, and, more recently, smart grid applications.

## Electricity measurement systems

The measurement system is the main subsystem of the metering system. It is composed of different components (hardware and software), which together are responsible for measuring, accumulating, and displaying energy in meters and instrument transformers. When associated with communication, data management, billing, and other systems discussed in this book, they form the metering system.

**Electromechanical induction meters.** The most common electromechanical meters are based on the Thompson design developed in 1888. An electromechanical induction meter is comprised of electrical and mechanical components that work together to register energy consumption. It may be single phase or polyphase. It is generally composed of the following components (see figs. 1–5 and 1–6):

1. Base and meter cover
2. Terminal connections and cover
3. Identification nameplate
4. Stator
5. Voltage circuit(s)
6. Current circuit(s)
7. Rotor disk(s)
8. Magnetic brake rotor
9. Pivot and spindle connected to a register
10. Calibration elements

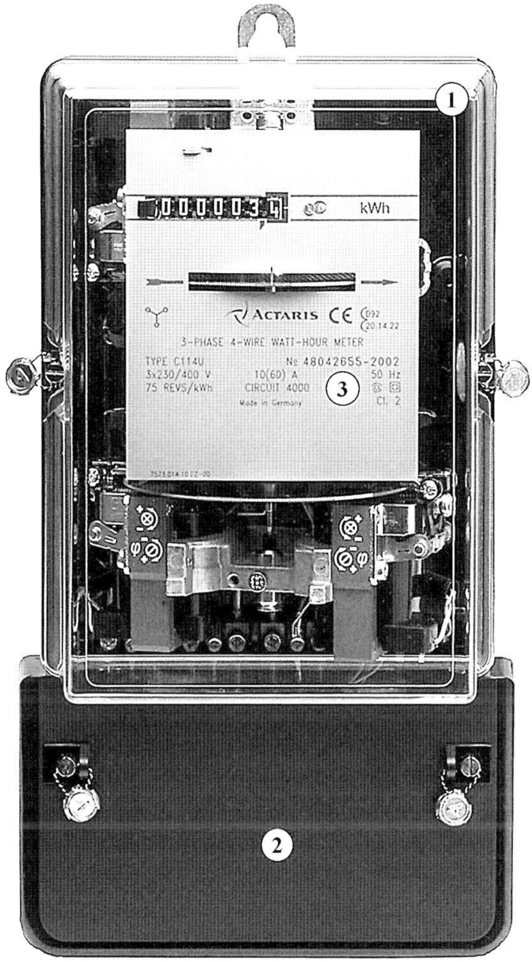


Fig. 1–5. Exterior components of an electromechanical meter. (Photo courtesy of Actaris.)

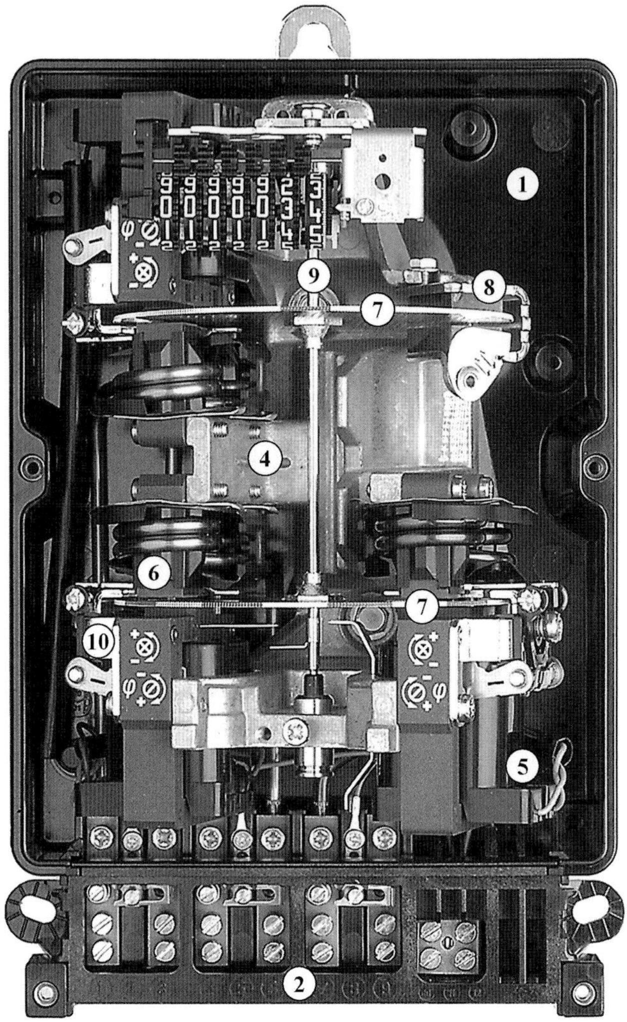


Fig. 1–6. Interior components of an electromechanical meter. (Photo courtesy of Actaris.)

A measurement element is composed of a voltage and current circuit. The voltage circuit contains a coil that is connected in parallel with the main circuit. The current circuit contains a current coil that is connected in series with the load. Polyphase meters have more than one measurement element, usually proportional to the number of phases. In general, single-phase meters have only one measurement element. However, special versions may have multiple elements. Single-phase multi-element meters are used when different circuits must be measured independently.

The voltage coil collects the voltage reference from the line, while the other reference is obtained from the current that flows within the current coil. Therefore the meter works in the same way as an induction engine. Electromagnetic fields are generated in proportion to the voltage supplied and to the current flowing through the meter. Together they result in the disk rotation.

The disk rotation causes the pivot to rotate, and its spindle rotates the gears of the register display in proportion to energy consumption. The gear system of the register display is composed of several cogs that are associated with subcounters that record total usage. Their number depends on the utility specifications and on the meter manufacturer design. Generally, the register display is composed of four, five, or six subcounters, each one corresponding to a digit. They are interconnected mechanically, and they may show decimal or unitary units. The register display may directly express consumption, or it may be necessary to multiply the values obtained from the register by a coefficient.

The magnetic brake rotor produces an opposite force to the disk rotation that controls its speed in relation to the energy used. It eliminates unnecessary acceleration or nonstop events associated with the inertia of the disk when the current increases or stops flowing. The calibration element screws are used to correct the accuracy of the meter if needed and are generally used during laboratory meter tests. The base and its cover are either glued or screwed together. Connection terminals are also within covers for safety reasons. Both meter and terminal covers are normally sealed to identify when tampering has occurred. A label is added to the meter to provide information about its identification, main features, and a connection diagram.

An electromechanical meter generally measures a single unit only. If it is necessary to measure kWh and kVArh for a customer, it may be necessary to employ two separate meters. For generating billing invoices, the utilities collect the registers from the meters over a defined period (for example, monthly). They compare the current values to those from the previous month in order to calculate the customer consumption.

**Programmable electronic registers.** Some electromechanical meters are designed to transmit pulses to other devices. They normally use a special infrared device for counting the disk turns and transmitting equivalent pulses to an embedded (fig. 1–7) or external source such as a programmable electronic register.

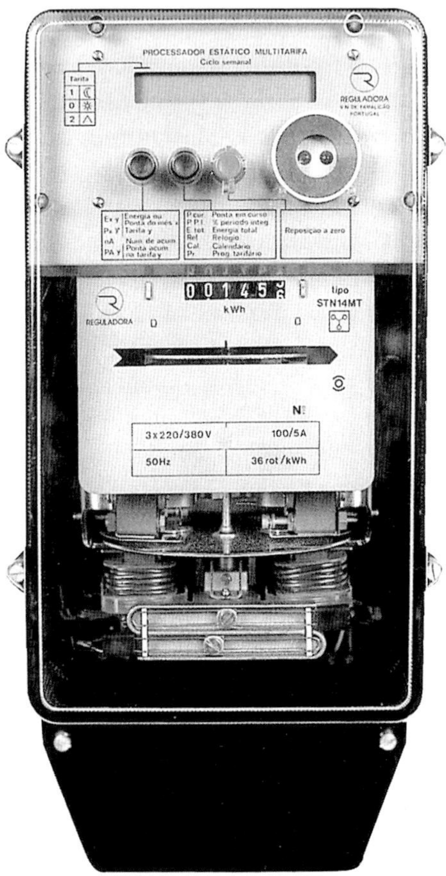


Fig. 1–7. Example of programmable electronic register. (Photo by Photo Ludo, courtesy of Actaris.)

This device is not considered a meter (it is not able to measure) but an electronic processor able to manage pulses from different emitters and store them in different channels. The equipment has an embedded clock, and its functions are programmable. It is designed to work in tandem with electromechanical meters for storing profile data and active demand usage, and also for providing hourly or seasonal rates or other clock- or calendar-dependent functions.

Some of these devices can also produce standardized pulse output, which may be used by energy management systems and devices owned by customers for demand management. Optionally, the basic pulses can be provided to the customer via an optical isolator that duplicates the pulses so they can be simultaneously used by the electronic register and the customer's management device.

Nowadays, even with the availability of electronic meters, programmable electronic registers are still common in utilities around the world. Traditionally, the measurement system used by these meters has been used for high-consumption customers based on hourly/seasonal rates. Often, in this scenario, active and reactive electromechanical meters provide pulses to a programmable electronic register. Note that this device is not energy totaling. It does not sum pulses generated by different meters. Instead, any received pulses are stored individually in different channels.

**Static electronic meters.** Once used only for high-consumption customers, electronic meters are now massively deployed in all kinds of facilities. They are used for the majority of smart metering installations in the world.

In contrast to programmable electronic registers, static electronic meters can measure energy, and they eliminate the need for any additional source of measurement by integrating all the functions of the electromechanical meters. Depending on specification, they can simultaneously measure different energy units such as kWh and kVArh as well as incoming and outgoing energy. They can provide the following energy management functions:

- Measurement of multiple energy units (energy, power, voltage, current, etc.)
- Rate calculations
- Operation of switches and contacts based on programmable functions and events
- Monitoring of events and sending alarms

- Load monitoring
- Telemetry functions
- Monetary conversions
- Consumption forecasts
- Energy measurement analysis
- Storage of profiles of energy and other measurement units

Electronic meters can manage single or multiple rates. Advanced hourly/seasonal rates, associated functions, and switching events may be provided via an internal clock or external sources, such as clock, broadcast radio frequency, ripple control, local data concentrators, and remote servers. These external sources are not restricted to providing clock references for rate period changes. They are also used to broadcast commands such as load shifting management for strategic loads during peak time and according to rate arrangements.

A static electronic meter is comprised of electrical and electronic components that work together to register energy consumption. This meter may be single phase or polyphase. It is generally composed of the following components (see fig. 1–8):

1. Base
2. Measurement board
3. Main printed circuit board
4. Identification nameplate
5. Main cover
6. Terminal connections and cover

Similar to those of electromechanical meters, the operating principles of static electronic meters are based on voltage and current measurements. These measurements are obtained with high precision by internal sensors located on the meter circuit board. These inputs are internally processed, and the different energy units are incremented and recorded according to standardized measurement algorithms.

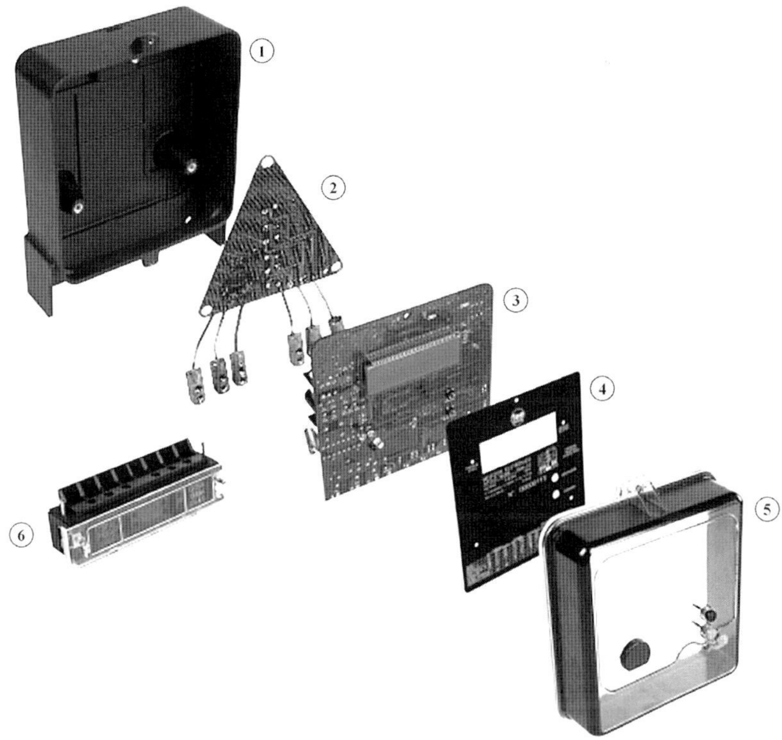


Fig. 1–8. Components of an electronic meter. (Photo courtesy of FAE.)

An interesting trend is the rise in importance of the electricity meter as more components and functions are integrated. For example, due to utility needs, in the majority of smart metering deployments in the world, a load switch may be considered an integral part of a basic smart meter. In other designs, integrated circuit breakers for safety operations or advanced functions are provided. Functions may relate to management of home automation, local data management for in-home devices, and dual-fuel arrangements (associated gas meter).

There are many different types of electronic meters. The traditional ones (fig. 1–9) are built in a single module with all components embedded and protected by a seal. If internal components need to be replaced (e.g., because of a switch fault), this can only be done in a laboratory environment. For metering accuracy and safety reasons, there are only a few accessible components available during normal operation, such as connection terminals and batteries.

Fig. 1–9. Example of a traditional meter used in France. (Photo Courtesy of Sagem.)

Another common design is the modular meter. These are initially more expensive, but may be advantageous for some metering architectures because they are upgradable in the field. Taking into account that different components have different life spans, a modular design allows the replacement of targeted components rather than the whole device. This replacement may occur for different reasons such as communication technology obsolescence, life span being reached, or accuracy verification.

The plug-in meter is an example of a modular meter; it is mainly deployed in South Africa (fig. 1–10). Here, the base and terminals can be unplugged from the metrological part of the meter and switch. This may be worthwhile, because the base components have a longer life span than the metrological part. It also improves meter installation time and associated costs.

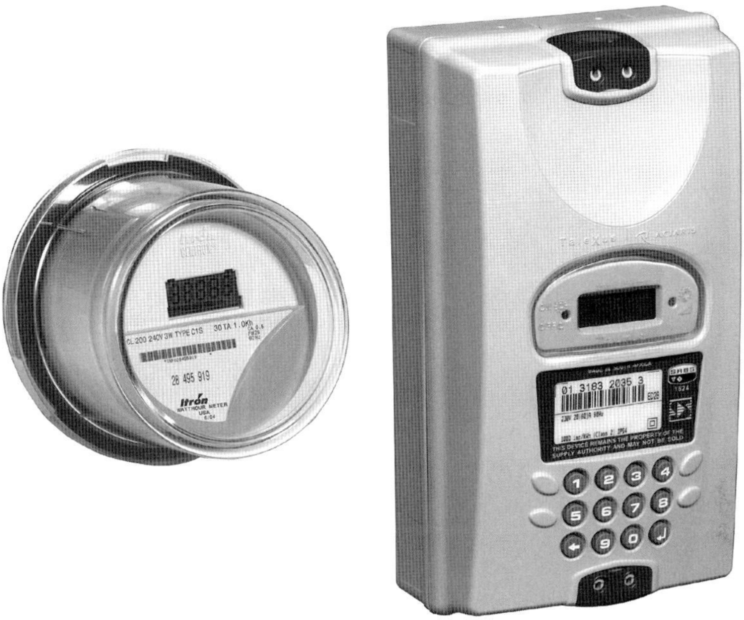

Fig. 1–10. Example of socket (left) and plug-in meters (right). (Photo by Itron and BMS Imaging, courtesy of Actaris.)

Socket meters are used in a similar way in North America (fig. 1–10). These differ only in base design and component locations.

Another interesting design is that of the rack meter, which is generally an indirect measurement system, used in panel arrangements for substations. This meter may be quickly and safely replaced in the field for different purposes. Its base normally has an internal mechanism to automatically short-circuit the secondary current and open the voltage circuit in order to isolate the meter prior to removal.

In general, modular meters offer more flexibility. They may allow the replacement of components in the field, such as communication modems and switches, through the use of fully isolated meter compartments. This may significantly reduce the cost of the field operations.

**Direct and indirect measurement systems.** Regarding the way they are connected to the grid, meters can be classified as direct or indirect measurement devices. Their application depends on the customer's load requirement. Direct measurement meters are usually applicable for residential customers, and indirect meters for bigger load customers.

Both types of meters have two measurement circuits to register energy: current and voltage. In a direct meter all the load of the circuit flows directly through its current circuit. The main supply cable provides both a voltage and current reference to register consumption. These circuits are often interconnected inside the meter (fig. 1–11). When needed, it is possible to separate them for laboratory test purposes. This is possible thanks to a specific connection link in these circuits.

In case the customer load exceeds the limits of a direct meter (typically up to 230/400 V and 100 A in some European countries), for example, in an industrial site, it is necessary to provide an indirect type of measurement. The meter adopted is thus an indirect measurement meter. Instrument transformers are required to both connect the indirect measurement meters to the grid and reduce the measured values to an acceptable limit that can be directly supported by these meters. In these measurement systems, voltage and current circuits are fully separated outside and inside the meter (fig. 1–12). Thus, these meters have two fully independent circuits, one for current and the other for voltage.

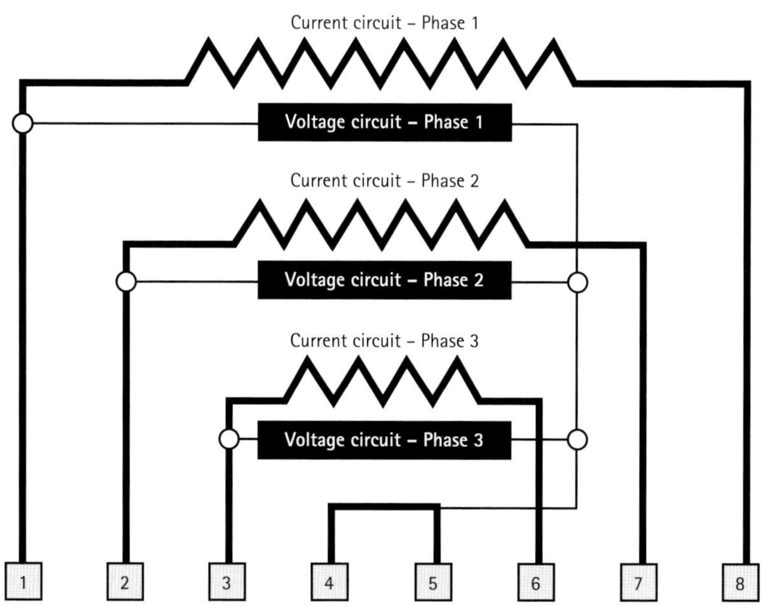


Fig. 1–11. Example of a direct meter circuit

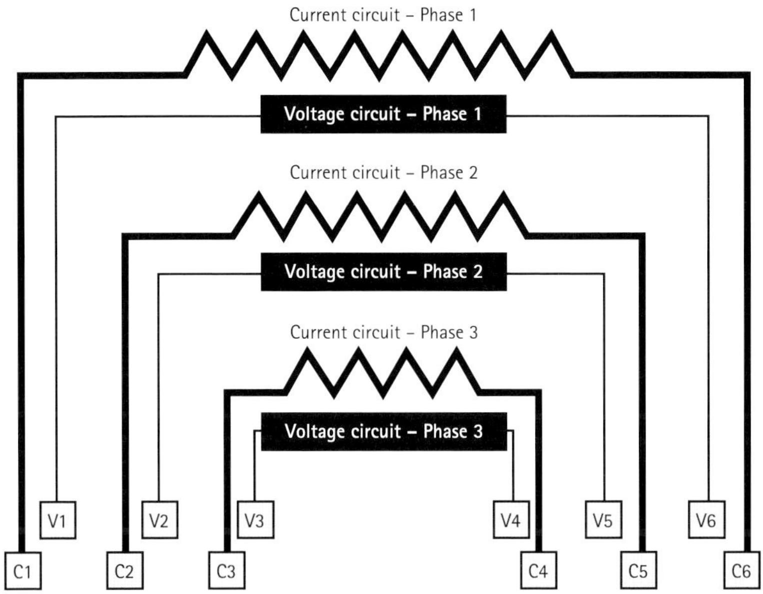


Fig. 1–12. Example of an indirect meter circuit.

***Instrument transformers.*** Depending on the customer load and the meter's features, different types of instrument transformers can be used to provide information on the electricity meter. There are two different instrument transformers applicable for measuring systems: current (fig. 1–13) and potential (fig. 1–14). They are associated with two circuits: a primary circuit that reflects the real supply values and a secondary circuit that reflects the transformed values for measurement purposes.

If the customer load can be met within the limits of a low-voltage feeder, it will only be necessary to install current transformers. The transformers reduce the current flowing on the primary circuit to an acceptable range of values to be measured. The typical range of current is 0–5 amperes for indirect meters.

Instrument transformers are associated with coefficients (multiplication factors) that reflect the transformation between their primary and secondary environments. These coefficients can be used in calculations to transform the measurements obtained by meters from secondary measurements to final values, which reflect the real consumption of the customer. Instrument transformers are associated with a rated burden in VA for safety reasons. The topology of measurement circuits differs from environment to environment and depends on the customer substation arrangements, manufacturers' specifications, and utility preferences.

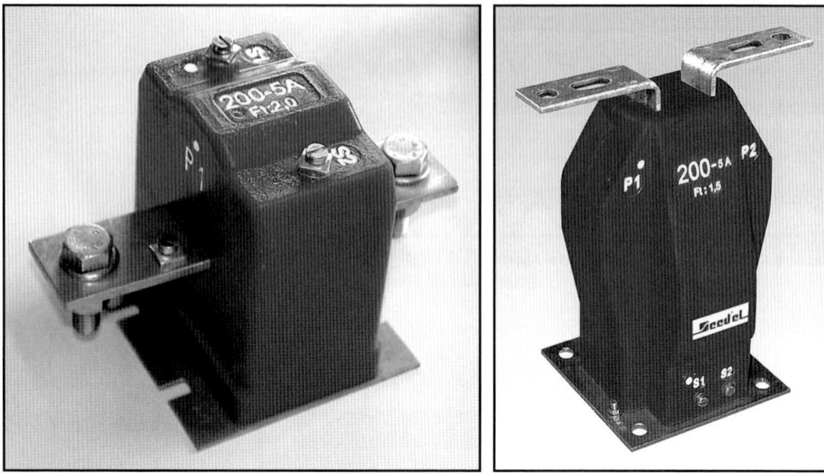


Fig. 1–13. Examples of low (left) and medium (right) voltage current transformers. (Photo courtesy of Seedel.)

Fig. 1–14. Examples of medium voltage potential transformers. (Photo courtesy of Seedel.)

Instrument transformers must be specified according to the features of the measuring system. For example, the ranges of secondary voltage and current vary according to utility needs. For specific applications, where the levels of current on the primary circuit are too small to be accurately measured by the meters, current transformers are used to increase the levels of current, instead of reducing them. The following are important parameters for instrument transformers:

- Primary and secondary voltage and current
- Number of secondary circuits
- Accuracy class
- Isolation class
- Rated burden (VA)
- Circuit topology
- Standards
- Any special requirements

Instrument transformers are generally installed in separate sealed compartments. These compartments may be placed in a cabinet or in a dedicated substation (common for medium-voltage-supplied customers). An important aspect on this kind of installation is its secondary circuit. Special precautions must be taken to clearly identify, secure, and isolate all cables. The same is applied to conduits not only for safety reasons but also to avoid mistakes and tampering. Conduits are used to accommodate the secondary circuits that connect instrument transformers to meters.

For accuracy verification in the field, appropriate test switches must be part of secondary circuits, which are considered to be in an uninterrupted loop while energized. Inappropriate open-circuit procedures may create safety risks and faults on the current transformers. Modern test switch equipment contains a semiautomatic mechanism used to short-circuit the secondary current transformers in order to isolate parts of the circuit where technicians need to work. This mechanism also opens the voltage circuits to isolate parts where test procedures will be executed. For safety reasons, all the technicians must be thoroughly and continuously trained for these procedures, as well as for any kind of procedure involving metering manipulations within energized environments. Safety must be paramount for metering specialists, managers, and the responsible companies.

Taking into account all the other equipment in the measurement circuit, the meter cannot be seen as the unique device associated with metrological procedures. It is just one of the components of the measurement system. Its instrument transformers and secondary circuits, and indeed all its components such as connections and test switches, are also part of this measurement system. Instrument transformers are precision instruments, and thus they must conform to accurate metrological standards and regulations. Metrological standards generally use parallelogram representation for defining accuracy limits.

***Compact indirect measurement systems.*** Manufacturers of instrument transformers take various needs into account such as accuracy, space constraints, safety, time required for installation, and reduction of nontechnical losses. In order to improve these factors, they have developed compact integrated systems for indirect measurements (fig. 1–15) that may be applicable for low or medium voltage, fed by (up to) 36-kV lines. Because of their easy installation, these systems are applicable for network, comparative, grid boundary, and other measurement purposes. This compact measurement system is comprised of an electricity meter, a customer display unit, current transformers, potential transformers (when applicable), and a protection circuit.

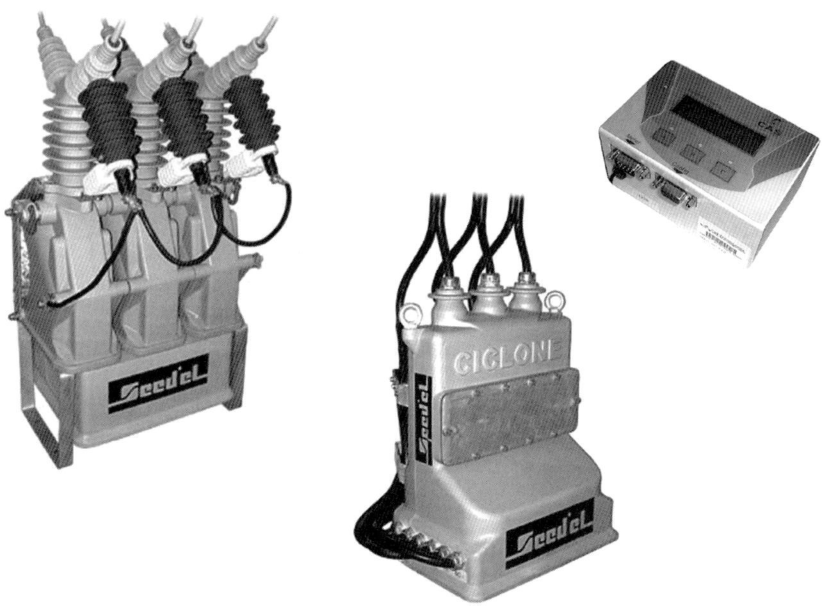


Fig. 1–15. A compact integrated system for indirect measurements with a customer display unit (right) for low- and medium-voltage (left) applications. (Photo courtesy of Seedel.)

Traditionally, every measurement device used in the measurement system is tested and calibrated separately. Given that the total error of the measurement system is the sum of all the individual errors, and taking into account that these components are made by different manufacturers, sometimes the total error of the system is significantly higher than its individual components. For compact integrated systems, if all the measurement equipment is produced by the same manufacturer, and the entire system is tested together as a single module, this kind of measurement system generally has greater accuracy.

As a natural commercial response to the market, meter manufacturers are now developing new direct-measurement meters for low-voltage use that are able to support electric current as high as 800 amperes for regular use (fig. 1–16).

Fig. 1–16. An 800-A meter. (Photo courtesy of Nansen.)

***Energy totaling (summating).*** Customers often have special requirements for totaling energy consumption. One example is a facility supplied by several meters where customers need a single-pulse signal for their local energy management device. Complexity may be increased if quality and reserve meters are added to this scenario for backup purposes. Another application of these systems is for submetering, which measures different parts of the customer's business and provides real-time data. In both cases, special methods must be used to summate the energy from multiple measurement points.

One method uses an electronic totalizing register. Its function is to sum the pulses or measurement data of different meters according to programmable functions and provide centralized, advanced, real-time management and analysis, such as a demand-exceeded signal. These devices are also used to generate signals to customers' energy management devices. Electronic totalizing registers may provide several additional functions such as the following:

- Global Positioning System (GPS) clock synchronization for local meters
- Centralized access to the meters locally or remotely
- Customer display units for advanced information
- Energy management

Another way of providing measurement aggregation in the field is to use a current addition method, which applies only to indirect measurement systems. This method enables physical association of several secondary current circuits in the field from different measurement points. These circuits are grouped and measured by a single meter common to them both. This may be done either directly or indirectly:

- Direct aggregation is applicable to current transformers with the same features. A common connection terminal joins similar phases of different related secondary circuits.
- Indirect aggregation is also used for similar current transformers but with different transformation rates. In this case, the connection between the different secondary circuits is provided through extra autotransformers.

In both cases all phase references of potential must come from the same transformer and be the same for all the circuits.

Although these systems are used by various utilities and may work properly, there are disadvantages, including installation constraints due to the distance between the instrument transformers and the common meter, technical losses, voltage or current inversions, different levels of voltage oscillations for the individual circuits, and difficult maintenance and field inspections. Because of these problems, these systems are generally being replaced by individual meters combined with electronic totalizing registers.

**Reactive energy measurement.** Electromechanical or electronic meters are able to measure reactive energy. For this purpose we generally use meters with features that are similar to those of meters used for measuring active energy. The main difference is in the measurement circuit where an angle shift is necessary between the current and the voltage.

For electromechanical meters, a special autotransformer is generally integrated into the circuit that is able to shift the electrical current and voltage by 90 degrees to enable the measurement of reactive energy. With this method the meter can be used for measurement purposes both when the circuit is capacitive and when it is inductive. Although theoretically such a meter would be able to count in both directions and accumulate positive or negative reactive energy (for inductive or capacitive circuits, respectively), in practice there is little interest in it doing this. For this reason, these meters usually employ a brake mechanism to prevent the disk turning in a nonregular way. For this reason the meter normally stops when the circuit is capacitive. This brake is necessary to avoid mixing, on a single register, the consumption of capacitive and inductive reactive energy. Thus, this method works properly for calculating rates associated with measuring inductive energy only. It is not efficient for rates that need to measure capacitive energy for a period of time and inductive energy for another.

To minimize this problem, some utilities use a process called method Q. The circuit is arranged in order to provide a shift of 60 degrees between the angles of the voltage and the current. Optionally, a special autotransformer may also be used to provide the same shifting effect. In both cases the meter measures a special unit called a kilo-quadergy hour (kQh). In a vectorial analysis, the kQh is shifted from the kVArh by 30 degrees (fig. 1–17). Using this measurement type, additional to the inductive energy, utility companies are able to measure capacitive energy for a power factor up to 0.866. For billing purposes, the conversion from kQh to kVArh and vice versa is done using specific calculations.

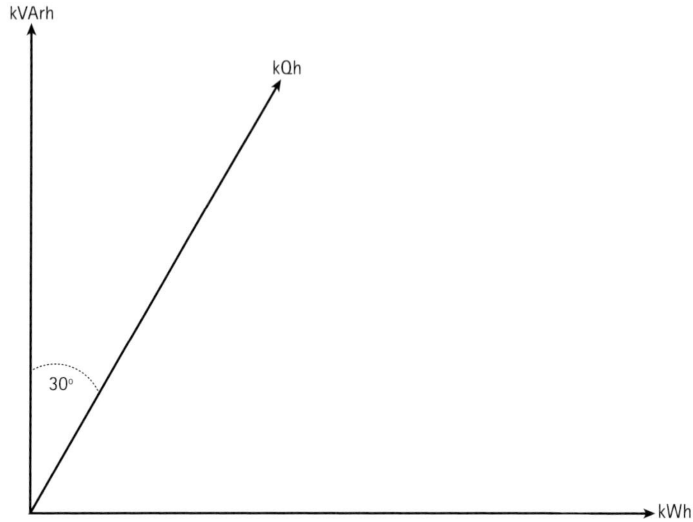


Fig. 1–17. kWh, kQh, and kVArh analysis

Although based on similar principles as those described earlier for electromechanical meters, thanks to the combination of high-precision sensors within electronic boards and efficient algorithms, electronic meters are able to measure reactive energy more efficiently in terms of cost, space, and performance. They also enable advanced functions such as time-calendar-based treatments and load profile recordings.

**Meter installation.** Because of its associated costs and technical impacts, installation is an important consideration in metering systems and must be carefully planned. The metering location and all the processes and resources related to its installation must be wisely chosen.

Safety recommendations must be respected during the installation procedure. Proper compartments have to be provided for mechanical protection for metering components (fig. 1–18), and they need to be properly sealed to prevent any kind of unauthorized access. A common example for electricity meters is the connection terminal, where cables must be properly secured with the correct screw arrangements and covers in order to avoid bad contacts in the connections. Safety must be paramount. Exhaustive training is important to mitigate the risks associated with metering procedures. A proper protection system is also necessary. For safety reasons, protection inside and outside the components of the metering systems must be provided. In general, several protection mechanisms are used for this such as electronic components, circuit breakers, and fuses.

Some utilities make use of existing meter boxes, mounting frames, or associated cabinets for installing these protection components.

Fig. 1–18. A meter installed in an exterior meter box. (Photo by Landis+Gyr.)

A proper commissioning procedure is also necessary to ensure that the meter is installed according to specifications and is operating correctly. This is also a good opportunity for safety inspections and to train customers on how to use functions available on their meters.

Metering systems can be installed either internally or externally to the customer premises. When installed internally, they are located in specific rooms, substations, cabinets, or wall panels. External installation is done for several reasons, depending on the different needs of the utilities and customers, such as reduction of operational costs for installation, maintenance and other field procedures, accessibility, and revenue protection. Metering systems are installed in individual or collective arrangements, grouped in the common areas of a building or in a specific room designed for this purpose only, or installed on distribution poles.

## Gas measurement systems

This section provides a high-level overview of gas meter operations. This is useful for understanding smart metering concepts. A gas meter is a device built to measure the flow of fuel gases. The measurement unit is generally cubic meters ($m^3$). While some utilities apply the cubic meters unit for billing purposes, others use defined conversion standards for transforming the flow to kWh units. Based on physical principles, temperature and gas pressure values are fundamental parameters for effective gas measurement. They are used to convert measurements obtained in the field into a billing standard. These conversions can be embedded within gas meters or calculated in the billing system.

Until recently gas meter design has been considered basic, but due to technology and smart metering applications, it has become more sophisticated. This has resulted in improved visual appearance and adaptations for space constraints when installing these devices in the field.

Electricity meters may have a load switch or a circuit breaker; the gas meter optionally has a valve. Valves are useful for services that depend on gas flow interruption such as remote valve operation and prepayment. Safety must be ensured for measurement and valve operation procedures. For example, when restoring gas flow, there must be no leaks or inadvertently opened valves. The meter or the metering system must provide mechanisms to ensure that this procedure is safely executed in the field. Examples are protection mechanisms, alarms, customer feedback functions, and the prompt interruption of the related flow when necessary.

There are several different types of gas meters used to measure flow:

- Diaphragm meters (fig. 1–19) are most commonly used for residential and some commercial applications, but were also used in the past for commercial and industrial applications. These are volumetric meters with diaphragms that expand or contract according to the gas flow. They can have a mechanical or electronic display.

- Rotary meters (fig. 1–20) are generally used for high-consumption applications. Basically, rotary meters use impellers that turn with the gas flow. They are volumetric meters and include two pistons, which, driven by the movement of fluid (gas) turn in opposite directions, staying in contact with each other and with the body wall of the meter. The rotation of pistons defines the volume of gas moved through the meter. The rotary movement is transmitted to a mechanical register or received by an electronic display.

Fig. 1–19. A diaphragm gas meter. (Photo courtesy of Actaris.)

Fig. 1–20. A rotary gas meter. (Photo by Michael Hec, courtesy of Actaris.)

- Turbine meters (fig. 1–21) basically measure the speed of a turbine within the meter forced by the gas flow. The turbine has a counter that registers consumption.
- Orifice meters use an orifice and differential pressure comparisons for measuring gas flow.
- Ultrasonic meters (fig. 1–22), as with turbine meters, measure the velocity of the gas flow within the pipe but use sensors and electronic components to perform complex calculations. The calculations are based on measuring the speed of ultrasonic waves propagated along the fluid flow; this difference is related to the velocity of the gas.
- Coriolis meters use calibrated pipes that vibrate according to the gas flow. Comparisons based on the frequency of these vibrations allow the measurement of consumption.

Fig. 1–21. A turbine gas meter. (Photo courtesy of Actaris.)

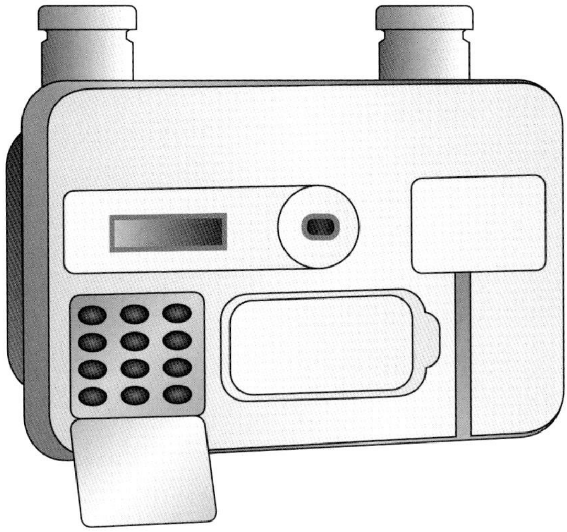

Fig. 1–22. An ultrasonic gas meter. (Courtesy of PRI.)

In terms of smart metering system, it is unlikely that the gas meter will add to its current functions of self-management and flow measurement. Internal DC batteries are necessary to supply electronic components of gas meters. This precludes adding extra online services. In the majority of premises where gas meters are installed, there are also electricity meters that are constantly AC supplied. In most smart metering architectures, the gas meter plays a secondary role. Its data are generally collected and managed by an electricity meter acting as an in-home metering gateway or by a separate, in-home gateway device. Even playing a secondary role, the gas meter is still an important element of dual-fuel systems.

## Dual-fuel metering systems

Many customers in the world are supplied by both gas and electricity, often provided by separate utility companies. In this scenario, customers may find that managing their energy consumption is complicated. In order to facilitate the energy management for dual-fuel premises, utilities may provide comprehensive energy services instead of single-fuel services. This offers a single measurement of energy used by customers.

Although the physical combination of both flow measurements into a single unit is only at the research stage, the logical combination of these elements is a reality in several countries in the world. Combined energy rates, services, and demand controls would benefit customers, utilities, and the environment. Customers are more likely to understand and manage their consumption if they receive combined energy information.

A common unit for gas and electricity meters is essential for dual-fuel measurement. Generally, the kilowatt hour (kWh) is used. Smart metering systems tend to convert this unit to a monetary value in order to facilitate energy management by customers. It helps them to budget necessary energy usage and can be displayed using a common display unit. Dual-fuel metering systems is further discussed in the next chapters of this book.

## Metering systems based on the use of customer display units

Customers are generally provided with details about their consumption in their energy bills. However, it is important for the customer to have a local and effective way of understanding their consumption. This applies to single-fuel metering systems, but is even more important for dual-fuel systems. Since meters can be read remotely and integrated with advanced IT systems, utilities can provide customer information many ways:

- Customer display units: Electronic devices associated with the customer metering system that can be installed in a home. Their level of complexity varies from utility to utility, from a basic consumption display to an advanced graphical display with in-home complex functions like temperature controls (fig. 1–23).

- Mobile phones: Some customers are using their mobile phones to access information about their consumption and new services. This is implemented in different ways, such as via a menu interface based on short message service (SMS), frequent reports, or a specific web page specially designed for mobile Internet access.

- Automatic call centers: Customers are able to call a specific number, generally free of charge, and get a report about their consumption.

- Automatic reports: Customers can be provided with predefined reports using e-mail, voice messages, and other methods of communication.

- Internet: A dedicated web page can be used to provide energy consumption information and associated services to customers.

- In-home devices: Appliances inside the home such as a television, an energy management system, or the customer's computer can be used for display purposes.

These displays create opportunities for utilities to interact with their customers. They are increasingly applicable to smart metering systems.

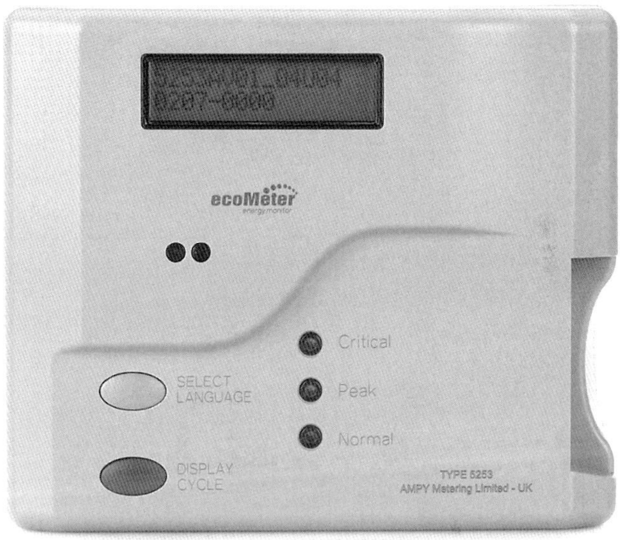


Fig. 1–23. A customer display unit. (Photo by Landis+Gyr.)

# Prepayment Systems

## Debit meters

Energy meters across the world may be technically classified according to their payment mode, either credit or debit. Credit mode was the first payment method implemented by utilities for billing purposes. It is basically a postpayment mode of energy consumption. By contrast, using the debit method, a credit is entered in the meter before energy consumption. The available credit is then decreased according to consumption. It is essentially a prepayment scheme, and depending on the country, either the term *debit meter* or *prepayment meter* is used.

The credit (or postpayment) method uses traditional meters for producing energy bills based on readings collected over a predefined cycle. The bill is then generated and paid by the customer. This method has several advantages, but also disadvantages; among them is the fact that the customers only know, via the utility, the amount of energy consumed at the end of the billing cycle. This can result in late or nonpayment, which is very difficult to manage between customers and utility companies. Nonpayment may result in disconnection of the customer supply by the utility. Poor management of energy—and associated budget problems—is one of the main reasons for supply disconnections.

Motivated to deal with this challenge, utilities have been increasingly producing packages of services and payment methods to be offered to their customers. They also developed methods of assisting their customers to efficiently manage their energy consumption and associated budget problems. The debit method is one possible solution. A debit meter may be created from a credit meter with an electricity load switch (or a valve in a gas meter) and extra processing for prepayment management purposes. Electricity debit meters are typically available in single-phase or polyphase direct-meter models. However, modern implementations include products for indirect meters.

With this method, as customers use their available energy credit, they can receive ongoing feedback from their meters. This means that they have a better understanding of their energy consumption, which in turn helps them to manage their consumption more efficiently. Several international cases demonstrate a significant reduction of energy consumption once these systems are deployed.

This flexible way of purchasing energy helps customers to adapt their budgets to their energy consumption and thus to pay their utility bills more easily. The bills may be divided into small, fixed payments to be processed at predefined times such as hourly, daily, or weekly. This method has been proven to be efficient; in international benchmarking exercises, credit purchase transactions are often done once or twice a week. There are also several cases of customers who purchase credits more than once a day. Traditional examples are fishers and recycling professionals in some countries who receive their own income weekly, daily, or even several times a day, and are thus more able to pay for their energy consumption in smaller and more frequent increments.

Based on the budget and energy management methods of prepayment systems, it is sometimes thought that prepayment is a service adapted only for customers with low incomes or for revenue-protection purposes.

However, international experience from countries such as the United Kingdom and South Africa shows this system is adaptable to anyone who would like to better manage their budget, consumption, or carbon emissions, or for customers with special energy consumption needs such as vacation homes, tenants, and students. Energy prepayment systems can also be thought of as similar to the prepaid services provided by many mobile telecom providers.

Since the end of the 19th century, a variety of prepayment products have been developed around the world. The first known prepayment device effectively deployed was the coin meter. After this, several types of debit meters were developed and implemented by utilities. Because of their wide variety, a benchmarking exercise is necessary in order to understand their features.

Now we can turn our attention to each technology and its associated features. Each has advantages and disadvantages, and each is more appropriate for a specific application. For this reason, this book does not suggest the best technology, but instead helps you to identify the most relevant technology for your needs.

## Classes of prepayment systems

The different prepayment technologies available in the world are mainly differentiated by the way the customers interact with them. These interactions happen principally when customers acquire credit, apply it to the meter, and use it. A prepayment system is comprised of not only the debit meter but also all its IT systems. They are deployed for many objectives such as customer management, credit generation, and credit management. Depending on the technology, more emphasis is given to the IT system or to the debit meter.

**Coin meters.** Many of the older coin meter models (fig. 1–24) were used in the United Kingdom at the end of the 19th century. Even now some utilities around the world continue to use these mechanically operated meters.

As the name implies, once the coins are inserted in the appropriate slot of the electricity or gas meter, it enables an amount of credit to be used. The coins are stored in a specific compartment inside the meter, and they are periodically collected by the utility.

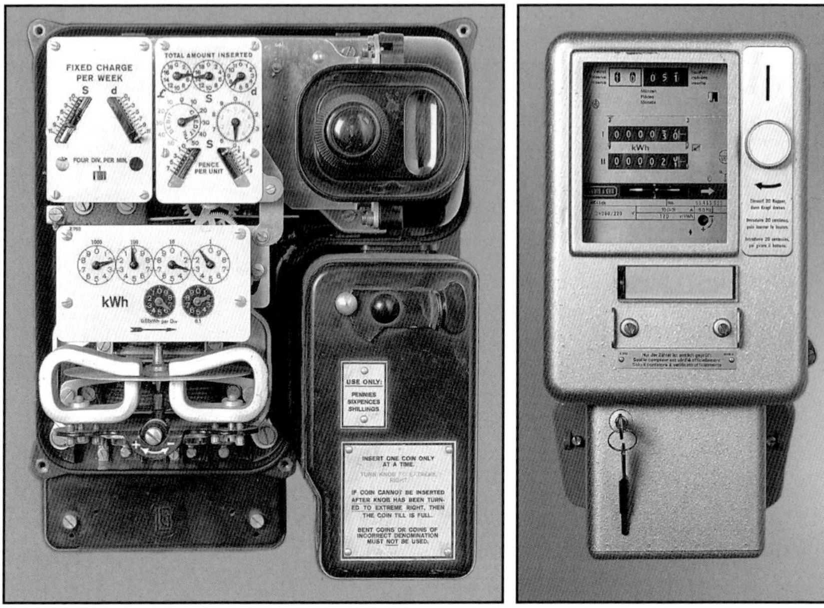


Fig. 1–24. An obsolete coin meter. (Photo by Landis+Gyr.)

In some countries, an interesting tendency was also observed. Coin meters were used for saving money temporarily, as a kind of safe, to be used when necessary in the future. This illustrates a trend utility companies could capitalize on—assisting their customers with budgeting for both energy expenses and for other needs.

The level of necessary IT systems associated with coin meters is very basic because most management is executed on the debit meter side. There is no need for selling credits, as the coins are a standard in the financial market. The main disadvantages observed by utilities are in operational procedures for collecting the coins in the field, tampering, vandalism, and theft. Another point is the complete absence of remote control or credit management.

Some people consider coin meters an obsolete technology, but that is not entirely true. New electronic modules, technologically upgraded, are currently available in the market as a new form of coin meters, timers, or sockets used by utilities or for private purposes. The coin meter remains a useful device for isolated private applications.

**Token meters.** This technology (fig. 1–25) is internationally known due to its application on payphones around the world. A similar technology has been implemented for energy meters and operates in a comparable mode. An energy credit is placed on a disposable card that can be read by the energy meter. It allows the load switch to be closed for the duration of the credit. The cards, known as tokens, may be associated with kilowatt hours or monetary values. The credit associated with these cards generally comes in two forms:

- Customized meter-specific credit: Pay points are supplied with card writer machines that can then encode the card with different amounts of credits that apply to a specific energy meter.
- Noncustomized and non-meter-specific credit: Sets of cards with predefined credits are generated and apply to any compatible meter within the utility. The range of available credits must be carefully studied because no customized credit purchase is allowed.

Fig. 1–25. A token meter. (Photo by Landis+Gyr.)

An advantage of preset value cards over customized cards is that customers can purchase these cards in different locations such as gas stations and supermarkets; no other control on the credit that is being used by a specific customer is required. Other features of prepayment token systems include the following:

- Simplicity of utilization and technology maturity
- Emergency credit available
- Relatively low-cost technology (despite the extra operational costs associated with the use of disposal cards)

In contrast to coin meters, disposable cards have created an interesting phenomenon. At the end of the credit period, when the credit on a disposable card is used up, collectors keep the cards, because, as with telephone cards, they have intrinsic value due to their different graphic designs. This opens up possibilities such as using metering devices or accessories for marketing purposes.

Generally, the IT payment system may manage the levels of credit purchased per customer and better understand related customer behaviors. This is clearly an evolution with respect to the coin meter class.

Even with this feature, the technology is classified as unidirectional. The credit is remotely generated (and optionally meter-specific), but no remote data retrieval or meter management is possible. For example, it is impossible to know if the credit was effectively applied to the meter without a visit in the field. Another challenge is the operational cost generated by changing meter parameters, such as rates.

The main market for this technology was in South Africa, although the cards have gradually been replaced by keypad technology and others meter types. An open standard for generation of these tokens has been defined by the standard transfer specification (STS) and is followed by several meter manufacturers and IT system providers. This standard has been converted to an International Electrotechnical Commission (IEC) standard. Other countries such as the United Kingdom and Australia have deployed token technology.

**Smart card meters.** Smart card meters (fig. 1–26) represent a natural evolution of the token meter. This technology is also internationally recognized and used in several areas, such as the restaurant industry. The technology works in a similar way to token meters, although the cards are reusable and are always meter-specific. For credit purchases, the customers must carry their smart cards with them.

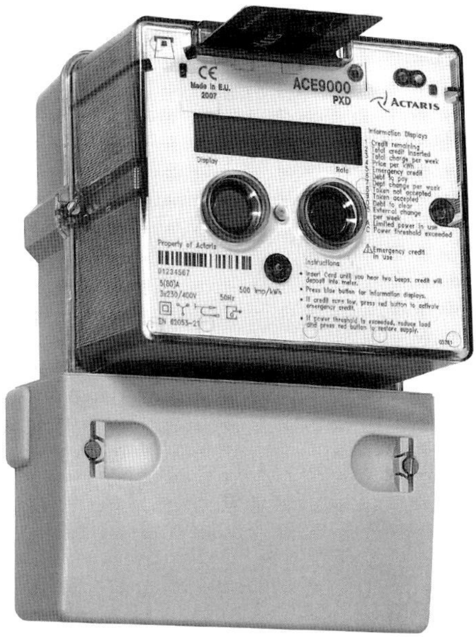


Fig. 1–26. Smart card meters. (Photo by BMS Imaging, courtesy of Actaris.)

An interesting development is that this card, in addition to sending the credit generated in the pay point to the meter, is also able to collect data from this device. When the card is inserted into the meter slot, it updates the customer's credit and simultaneously stores, on the card, metering data such as energy consumption, alerts, or even energy profile data. When the customer purchases more credit and it is added to the card, the metering data are collected and transferred via the IT system to a central database to be analyzed. The data can also be transferred from the IT system to the meters when changing parameters such as rate. Because of these features, this technology is called two-way technology.

Extra features include the following:

- Extra protection against fraud (smart cards)
- Local debt and fixed-charge management
- Disconnection profile and multirates associated with time, date, and emergency credit

This technology is mature and currently used for energy suppliers in different countries in Europe and Africa.

One of the technology limitations is the frequency of data update. The periods of data collection or parameterization depend on customer behavior in terms of credit purchase. Based on this, the data synchronization is the time elapsed between two credit purchases. Other limitations are smart cards management and customer dependence on the pay points to purchase credit. Problems may also occur with card management, such as card errors or lost or stolen cards, and must be promptly solved in order to preserve a good relationship with customers. It is also important to remember that, in general, technologies related to card readers are highly exposed to vandalism.

**Key meters.** Key meters (fig. 1–27) use technology similar to that of smart cards but customized for the energy sector. Key meters have been available on the market since around 1990. Instead of a card, they use a key embedded with a chip that stores and transfers meter data in the same way as a smart card.

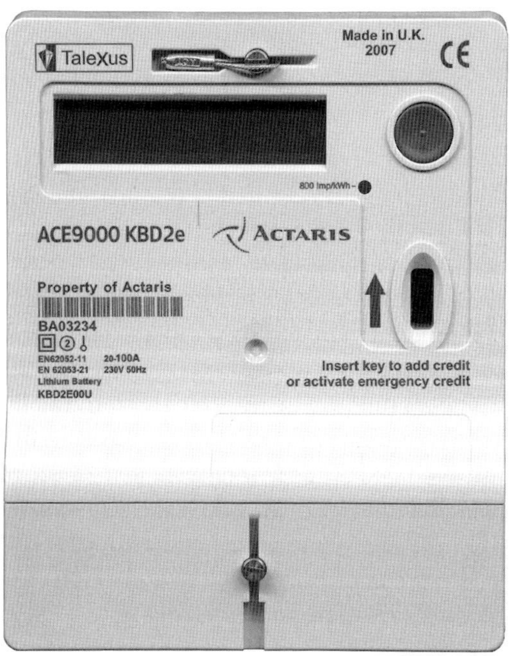


Fig. 1–27. A key meter. (Photo by BMS Imaging, courtesy of Actaris.)

International benchmarking exercises show that competitive energy markets may produce additional challenges for this technology. Multiple energy suppliers' arrangements may entail extra costs because of sophisticated IT systems and operational procedures for credit management and financial reconciliation, assuming a common pay point infrastructure shared with different utilities.

Difficulties may arise because customers change their energy providers from time to time, and often they keep their old keys (from the previous supplier). However, if a customer uses an old key to purchase electricity from a new supplier, the credit will be incorrectly associated with the previous supplier instead of the current one, and the credit must be reallocated afterwards. Despite this complication, this technology was successfully deployed, in large scale, in the competitive energy market in the United Kingdom and in relatively low scale in other countries.

**Keypad meters.** In this system, meters are equipped with a keypad as a substitute for a card reader or coin slot. The keypad is used to download credit or parameters to the meter. These credits and parameters are transferred to the meters via encrypted codes known as tokens. When customers purchase credit, they receive a code. When typed into the keypad, the meter decrypts the code and updates its credit level. This technology is classified as one-way, although extra devices are available to transform it to two-way. These devices can communicate via local or remote communication media, such as low power RF (radio frequency) or general packet radio service (GPRS).

Comparable to token meters, keypad meters use an open standard to define the codes. This standard, defined by STS and IEC, defines the algorithms and rules for the prepayment environment and is followed by several meter manufacturers and IT solution providers. The credits, which are meter- and utility-specific and are used just once, are transferred to the meters in kilowatt hours. Any calculations for rate conversion and payments such as taxes, fixed charges, debts, and extra services are performed in the IT system. As a result of market or metering supply choices, other standards are used for producing similar codes.

In operational terms, with credit purchase, the customer provides his or her meter number and pays an amount of credit. Once the IT system receives the payment amount and meter number, the value is credited to the customer's account, and the net credit is converted to a kWh credit. A code is delivered to the customers, which is then entered on the keypad meter.

Because it does not depend on any card or key to transport credit or related pay points, this is a very flexible technology in terms of sales channels. This is an important feature that permits the purchase of credit remotely, additional to the pay point option. Examples of remote channels deployed around the world are the following:

- Scratch cards: These cards are available in predefined monetary values at different sale points such as gas stations, supermarkets, and utilities. Before being attributed to a specific meter by the customer, the credit is meter-neutral. After purchasing a scratch card, the customer may associate it to a specific meter. For this, the customer must scratch the card and provide both the secret code and meter number. This can be via SMS using a predefined message format, the Internet, or by dialing into a call center. The monetary value, at this stage, is transformed into a meter-specific kWh credit. A corresponding code is delivered to the customer, automatically using SMS, web, or phone format. The customer then types it into the keypad meter.

- Wireless application protocol (WAP) or Internet: Credit card information is stored in the IT system, and then an identification number and password are given to the customers, who may thereby purchase credit remotely.

- Automatic profile: Similar to WAP, the credit card information is stored in the IT system, but the customer sets a quantity of credit to be automatically purchased at set intervals, such as once a week. The credit code is sent to the customer via different media, such as secure links using e-mail or SMS.

- Automatic call center: Similar to automatic profile, but using an automatic call center to provide the customer with the code by voice message on the phone.

- Automatic teller machine (ATM): Special terminals allow the customer to provide a meter number and pay for energy by cash or card.

- GPRS terminals: These devices are installed in sale points such as gas stations for credit purchase.

- Virtual private network (VPN): A VPN is installed in places such as supermarkets by the energy supplier and a partner; the credit can be purchased by customers in these locations.

Keypad meters are available in three different designs:

- Traditional single-phase or polyphase meters are built as a single module (fig. 1–28).
- Plug-in meters are installed inside the customer's home and provide easy maintenance. These meters have two parts:
    - Base: the wires and terminals
    - Measuring device: the meter and switch or circuit breaker
- Split meters have a display separate from the meter itself. The meter is then installed outside the customer premises such as in special boxes on distribution poles or on the ground. The customer display unit is installed inside the customer's home. Split meters may be interconnected by a wired network, low-power RF, or power line carrier (PLC). Among several other applications, the split meter may be an interesting option for revenue protection actions.

Fig. 1–28. A traditional keypad meter. (Photo by Landis+Gyr.)

At this time, South Africa is the largest market in the world for keypad meters. They are also deployed in several locations in the world including South America, Europe, and Oceania.

**Remotely managed meters.** The remotely managed meter class represents a natural evolution from the previously discussed types. This meter is basically a debit meter with a wide area network (WAN) communication device, although it can use different communication techniques.

In contrast to the technologies that require tokens, cards, or keys for data transportation, this prepayment system requires no extra action from customers. As in the keypad model, the customer is able to purchase credit at a distance via different sales channels. Taking into account the WAN communication capability of the meters, once the customer purchases electricity, the credit is automatically transferred from the IT system and updated at the meter. The same procedure applies to updating parameters.

A remotely managed meter can be provided with a physical keypad (fig. 1–29) or a virtual keypad controlled by buttons. In order to deal with communication failures, as with keypad meters, some systems provide 20-digit codes to use for credit purchase, as a backup method. If communication fails, customers may optionally type the code on the keypad meter, and the credit will be locally updated.

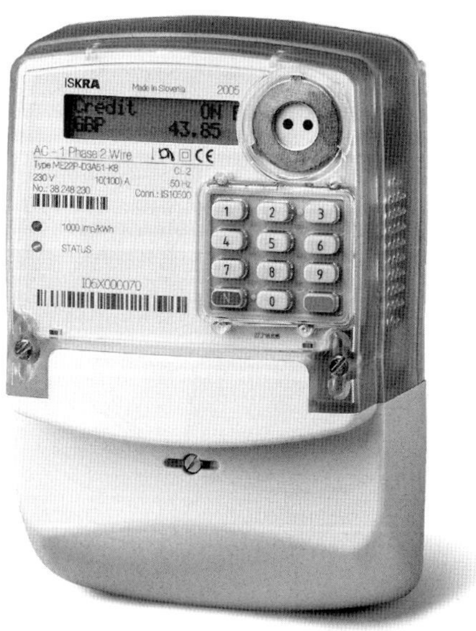


Fig. 1–29. A remotely managed meter. (Photo courtesy of Iskraemeco.)

Credit and other prepayment data management can be accomplished two different ways:

- Locally in the debit meter: As with key meters or smart card meters, the monetary credit is locally managed, and any consequential taxes or fees are managed by the meter itself.
- Remotely in the IT system: Meters are connected to the server in an online mode. Credit management and any other data management are executed remotely by the IT system and frequently displayed in the meter. A typical credit meter may be also used for offering prepayment services, with all calculations and management centralized in the IT system.

Several technologies are being used to provide this kind of prepayment system, such as SMS or GPRS meters. They are being implemented in many countries. This technology provides a very advanced level of prepayment services. This kind of technology, its potential services, and other future-proofing services are essential in smart metering systems.

## Emergency credit and social disconnection

Prepayment systems are recognized as a way to avoid local interventions for energy supply disconnection, as debit meters can self-disconnect the energy supply. But, as with any other service, it can always be improved. In order to help their consumers with financial difficulties, research and development departments of utilities and solution providers are working toward creating new services and opportunities for customers.

Emergency credit was the first known initiative in this direction. When the meter credit reaches zero, customers may activate an emergency credit. In other words, an emergency credit level can be added once the previous one has been exhausted. This is common for avoiding disconnections during nights, weekends, or holidays. Similarly, some energy suppliers around the world, often subsidized by local governments, offer monthly free energy credits to their customers.

This solves some problems, but can also generate new debt to consumers that will have to be managed in the future. The energy supply may still be cut off once the emergency credit ends. Some utilities in the world have implemented social disconnection services to deal with this challenge. With the social disconnection method, an initial emergency credit is offered to customers. While this credit lasts, energy demand is fully available for the customers' consumption, without any power limitation. When this credit limit is reached, instead of disconnection, new emergency credits are offered, but they are associated with power limitation.

These limitations tend to reduce the power supplied to a premise to a level that provides only basic needs. Local government authorities usually determine those amounts. As soon as a predefined amount of credit is used, the limitations come into effect. In this way, both a full disconnection event and the creation of nonpayable debts are avoided. After a defined period under minimum power limitation, the consumer may be fully disconnected or subsidized by government or charity associations.

In order to better understand the concept, consider the following example: A premise is regularly supplied with less than 6 kW of maximum allowed power. The first emergency credit amount may be associated with this value. When it is reached, another amount is offered, but now associated with a power limitation of 3 kW. At its end, this level could be now reduced to just 1 kW. After a predefined time elapses, this customer can be optionally fully disconnected. In this scenario, a more flexible way of providing emergency credit is associated with monetary amounts, power limits, and predefined time periods.

## Energy package rates

Some utilities are testing products related to an energy package rate. This kind of package is similar to that of mobile telecom providers in which customers may contract monthly packages for Internet use, calls, and SMS. This system works in different modes:

- Advisor mode: Customers establish usage targets based on their number of credits contracted for a predefined period. For example, a customer who contracts for 900 kWh per month could set a flat profile of 30 kWh per day. Once the daily value is reached, customers are informed via alert events. At this stage, they are able to analyze their consumption and adapt their behavior if necessary. This profile may be more sophisticated in order to include features such as hourly/seasonal profiles.

- Curtailment mode: Customers are given a power limitation profile. They are temporarily disconnected when a specific amount of energy consumption is achieved within a predefined time. This cutoff can take only a few seconds. This mode is a kind of a drip-in system. For example, customers may contract for a monthly package of energy and divide their available credit into small blocks of energy. Let us assume here a flat division in amounts of 10 watt hours (Wh) per minute. In this case the meter would authorize an average use of 10 Wh every minute. If this amount is not fully used, it will accumulate in the meter for the next period. When this accumulated amount is exhausted, the meter is disconnected until a new energy block is available. In this scenario, consumers would wait for one minute. After this time elapses, the meter is safely allowed to reconnect and permits the use of the next credit block.

One of the several advantages of these services is that the customers often know how much they are supposed to pay for energy in a period of time. This may help them learn how to manage their consumption. This may prevent problems of debt and delinquent payment. This service is normally associated with a prepayment service. Thus, if the customers decide to use more energy in a specific period than their energy package allows, they can just add a new credit to their meters. This avoids renegotiating their contract arrangements for temporary requirements.

# Telemetry and Telecommand

Telemetry and telecommand methods allow device data collection, parameterization, and controls to be remotely executed. In the metering domain, these techniques can be classified in two categories. The first uses a local area network (LAN) technology to provide remote access to the devices within a short range. This was the first technology that was available in the market. The second uses an additional WAN technology to provide access fully remotely.

## Short-range techniques

**Concentrated access points.** This was probably the first network to be used for metering purposes. It normally uses a wired LAN. Meters are interconnected via a defined medium, based on a specific standard. A locally concentrated access point is available in order to provide individual access to them. These networks are limited to a fixed number of devices.

There are several recognized standards available in the market and used by manufacturers and solution providers of metering systems. Examples of this kind of network standard are the Meter-Bus (M-Bus) and the EURIDIS.

The concentrated local access point can use different technologies, such as a magnetic plug or a traditional serial port. Energy management or display devices could be connected to it in order to provide a centralized view of the devices in this network. They execute advanced analysis such as local demand management or energy balance for transformers. Handheld units can also be connected to these ports and may be used for centralized data collection or parameterization purposes.

The disadvantages of these arrangements are the networks' operational costs for installation and maintenance and the need to regularly send technicians onsite for data collection and parameterization purposes. The time elapsed from an event recorded and its data collection in the field may also be too long and produce undesirable delays. A clear example of this is reporting a meter fault or a tamper. A vast number of tampers occur by manipulation events (devices are constantly installed and removed), and by the time an inspection takes place in the field, the fraud device may have already been removed.

An advantage is the reduction of operational costs, for example, because utilities no longer need to access their meters individually. These costs become significant when meters are installed in customers' premises.

**Independent access points.** Unlike concentrated access points, these local networks do not require any centralized point of access. A classic example of independent access points is found in short-range, low-power RF networks. In this case, handheld units and meters are both equipped with an RF module that is used for data transfer events. Meters may then be individually managed.

Interesting services may be employed, such as data transfer events using walk-by or drive-by services. A handheld unit is connected to a special RF master and is embedded with software that is responsible for managing local automatic data collection and parameterization. This device is given to a technician (fig. 1–30), who walks close to the premises (walk-by). It can also be installed in a vehicle that is driven next to the targeted consumers (drive-by). The available range of the communication technology used defines how close it must be to the meters.

When the meters are detected, they automatically send their metering data such as registers and alarms to the handheld units. Simultaneously, these devices are updated with new parameters, like change of rates, if required. When the vehicles or the technicians return from the field to their utilities, the collected data are then transferred to a central database for further processing.

Fig. 1–30. A special handheld unit used for walk-by services. (Photo Courtesy of Actaris.)

In the drive-by mode, it is very common to employ public service vehicles such as refuse collection trucks. These are constantly driven around a specific region, and given that the data transfer service is fully automatic, the extra operational cost required is minimal. The time required for data gathering is relatively short. Similar to the concentrated access point method, one of the disadvantages of this system, is related to the time elapsed (even if reduced) between the event record and its collection from the field.

## Wide area techniques

Because of several needs, such as those associated with more frequent remote data transfer events, short-range techniques were upgraded with WAN devices and complex management systems. There are many terms associated with methods of remote data transfer and management, and new terms are created every day. Specialists still disagree about what exactly these terms mean and their boundaries. It is more important, however, to understand the opportunities and challenges of wide area techniques. All their features and functionalities combined are part of the technology used in smart metering implementations.

Among these methods, the first that appeared on the market was automatic meter reading (AMR), a technique to remotely collect metering data and associated parameters. After AMR, automatic meter management (AMM) came on the scene. It includes the functionalities of AMR deployments and offers extra capability for management of data, services, and devices. Metering data are recognized and prioritized in order to generate benefits to those in the energy market. Bidirectional communication is essential for AMM technologies.

Another concept is advanced metering infrastructure (AMI), which contains all the functionalities of AMR and AMM implementations and also addresses some other areas, such as interoperability, data security, scalability, and future-proofing integrations. AMI is considered the most complete technology for advanced metering systems.

In terms of data transfer, wide area techniques can be organized in two ways:

- Individual device management: Using this method each device is connected to an individual WAN modem. Thus, they are able to communicate directly with the IT system without any other auxiliary device. An example of a device using this technology is a GPRS meter.

- Concentrated device management: Devices are grouped in the field via data concentrators. They are able to centralize and manage a number of points in the field and link them to the WAN. They provide the bridge between the devices and their respective IT system. PLC is an example of this technology.

## Data management and integration

An important concept for metering systems is data management. Different devices are able to remotely communicate with IT systems. However, they are only able to deliver benefits to utilities if associated with an advanced IT metering system that efficiently analyzes the data and consequently produces strategic information to several participants within the energy market, such as customers, regulators, energy service providers, and utilities. Data integration is another essential aspect because sometimes the data can only become valuable information if associated with other data sources like external databases and systems. An example is the association between data from the energy network platforms and metering systems.

Many systems interact with metering systems or depend on them. Thus, it is important to have a platform for all data management transactions. The platform is generally responsible for managing the metering devices in the field and their data. It is also responsible for data collection, data integration, and events like alarms and logs.

The most important IT system to be integrated with metering platforms is billing, which plays a key role in revenue collection for utilities. In metering terms, billing is responsible for linking the registers (originated by the meters installed in the field) to complex rates and other billing parameters such as taxes and energy prices. It is also responsible for generating and managing utility bills and payments. Meter information can be easily shown on utility bills. Historically, these bills are one of the main channels of relationship between customers and suppliers.

It is essential that the meters and the billing system are fully synchronized. Problems of coordination between them are frequently responsible for revenue losses within energy utilities. Other systems can be integrated with metering platforms, such as payment and energy network management applications.

In the smart metering world, the IT metering platform is crucial. Several features can be incorporated in this system because of its key role in management and integration. This system is known as middleware. Middleware is a primary focus of smart metering implementations.

In practice, the most important component related to a metering system is IT. It is very important to optimize integration among IT systems, meters, and associated infrastructures in the field. Historically, specialists from the metering department were responsible for managing their isolated and dedicated metering software, while specialists from the IT department were responsible for managing the complex corporate systems. This kind of structure worked properly until metering became more strategic and its systems more complex. Now, these two groups of specialists work together in order to provide an optimized platform to the utilities and the market. The difficulty is, as they did not work together in the past, metering and IT departments still tend to work in an isolated way. The relationship between these two groups is sometimes nonexistent and difficult to improve.

The opportunities around metering system implementations can only be met if both groups work in a fully integrated way. It is important to realize that this new environment may require some corporate reorganization, governance structures, training, efficient project management, communication, and other advancements.

# Revenue Protection

Revenue protection means reducing losses, whether they are classified as technical or nontechnical. Technical losses are involuntary, naturally generated, and related to the energy environment, such as losses produced by the Joule effect on electrical cable. Nontechnical losses may be voluntary or involuntary and due to mistakes, excessive delays to treat faults, or illicit actions; in other words, energy effectively delivered to the customers but not properly billed.

This section provides an overview of losses because one of the most important motivations for international deployment of projects involving metering systems is to prevent loss. Financial losses are a challenge for many utilities in the world, their customers, and other participants in the energy market. There are three causes for losses:

- **Mistakes** may happen in the field, in work done by metering specialists, or in the IT system within the customer or billing database. They are involuntary but may generate losses when they occur. Examples are inversion of cable polarity during a meter installation process, an inaccurate collection of registers, and an active customer in the field but not shown in the billing system.

- **Faults** are normal operational problems that may occur in several components of the metering system, such as an instrument transformer.

- **Tampering** is deliberate action in the field or corporate metering environment to generate reduction in customer billing. Examples are meter bypasses or illicit change of factors related to indirect meters within the billing system.

The most essential part of a revenue protection service is its professional staff. Based on the required technologies, expertise, level of actions in various sectors, and the investment necessary to achieve the utility targets, a project should be specially set up for this purpose. Smart metering projects seem to represent key opportunities for utilities that would like to combat their nontechnical losses.

Smart meters are essential in assisting utility companies both to achieve their revenue protection goals and create new opportunities for their customers.

## Technical losses

Technical losses can be produced by several components of an energy network and, in metering systems, apply only to electricity meters. Gas meters have to be battery powered, whereas electricity meters are, of course, powered by the grid. The energy that electricity metering systems themselves use is considered meter self-consumption and the consumption of its associated devices such as instrument transformers.

Regarding operational aspects, however, technical losses apply to both gas and electricity metering systems due to associated auxiliary components, for example, a WAN modem.

Meter burdens are generally based on international and local standards, but efforts are made to reduce losses. Manufacturers are designing these devices with more and more aggressive loss goals. Utilities must also specify their standards and specifications in order to meet their different objectives related to revenue protection, energy consumption reduction, and carbon emission targets.

These losses are sometimes ignored by some utilities due to the low levels of consumption in a single metering system. However, taking account of other factors such as the entire life span of these platforms and the total number of installations, technical losses may become a significant amount of energy consumption and consequent revenue loss.

For illustration purposes, let's consider a residential park of single-phase electricity meters without any auxiliary devices. The following is additional information:

| | |
|---|---|
| Quantity of meters | 5 million |
| Total individual meter consumption | 1 watt |
| Life span per meter | 20 years (about 175,000 hours) |

Assuming that meters are installed for their full life span, an individual meter will consume around 175 kWh, and the entire metering park 850 GWh in this period. Considering monetary conversions and financial aspects, a significant amount of revenue loss to the utilities could result.

Statistical reports, energy balance in the energy network, and preventive maintenance are important tools for utilities to use to reduce their level of technical losses. Actions taken to reduce nontechnical losses may also influence technical ones, such as changing the electrical cable structure to reduce energy theft, which will also result in an improvement in the quality of the grid.

## Nontechnical losses

Several factors influence nontechnical losses in many countries. These losses are sometimes controlled, and sometimes they are not. They affect the utility practical processes in an end-to-end way. These factors are generally related to the following:

- Inefficient utility processes: unbilled energy, meter data collection, invoice delivery and collection, bills paid late, and unpaid bills
- Utility environment and market behaviors: cultural, legislation, regulation, and social

Considering these factors, savings will not be achieved if technical actions in the field are disconnected from the environment and market behavior aspects. Sometimes, cultural actions and education campaigns are necessary for changing established views. For example, if either due to impunity or lack of information, some consumers believe that energy theft is not a crime. Some utilities encourage conservation by distributing electricity-saving light bulbs and bonus vouchers to purchase energy-efficient appliances. These programs demonstrate the importance of optimizing energy consumption and help to lower costs for low-income consumers.

Nontechnical loss levels are also based on fully dynamic conditions, and because of this, continuous actions need to be implemented in several sectors. It is important to prevent and reduce the level of losses, but it is also essential to educate customers on how to use their energy efficiently. However, utilities still need to assist customers with their budget and payment behaviors with respect to their consumption.

An end-to-end utility process map and analysis are recommended. All processes should be carefully analyzed because sometimes indirect processes may affect utility losses, for example, fault-related losses. Devices can develop faults, and metering system components are no exception to this rule. Faults in metering system components may produce incorrect measurement of energy consumption. As these are hard to identify, they eventually stay in the field for a long time until identified and corrected. Immediate detection of and solutions for problems in the field may significantly reduce the level of losses.

Another interesting example is related to public streetlights. In some countries they are billed by utilities based on average forecasts, although several factors may influence these forecasts and they are unlikely to be precise. Seasonal parameters affect daylight time and respective energy consumption. It is also hard to synchronize the utility database with the ones from public lighting service providers. Uncontrolled installations, maintenance, and upgrades may occur. Average payment structures are common. This method of billing does not stimulate reduced levels of losses in their networks. Technical losses may additionally occur if they use private circuits or inefficient equipment, which is then unbilled. With nontechnical losses it is no different; clandestine connections may be made in these streetlight circuits that are hard to find.

Technical action is necessary to assist utilities to reduce their levels of nontechnical losses. A nontechnical loss is a complex subject and a growing challenge for some utilities in the world.

**Intelligent analysis.** Administrative mistakes are a common source of losses for utilities. The majority of them are in utility database registers and controls. Unmeasured energy sometimes comes from clandestine connections, but it is also related to database mistakes or inefficient updates. The inputs and outputs of these databases are other sources of losses when badly managed. Incorrect meter data collection, invoice delivery, and billing collection are common examples.

One of the first actions that utilities should take is to analyze and update their databases. Analysis tools are important platforms to be used in revenue protection. Another important aspect is the change of organizational structure. For utilities with a serious problem of nontechnical losses, it is essential to create an intelligence center exclusively for this purpose. It must provide dynamic controls on the losses and feedback on actions to reduce them. It is important that they will be able to control the level of losses for different network boundaries, such as grid transformers and feeders for electricity grids. This may be achieved with the installation of meters at strategic points for energy balance (fig. 1–31). Due to the dynamics of the losses, these analyses should be graphical, based on a profile. Periodical balances that are not based on load profiles are inefficient for revenue protection actions. For example, a level of 10% of nontechnical losses in a month may have no bearing on revenue protection actions. This percentage may reflect different events such as tampers or faults that are still in place or were detected and already corrected.

The center should be provided with a great level of detail on metering such as logs, alarms, and advanced information, such as phase vectors for electricity meters. As much of the analysis provided by this center as possible should be online to improve efficiency of the revenue protection service.

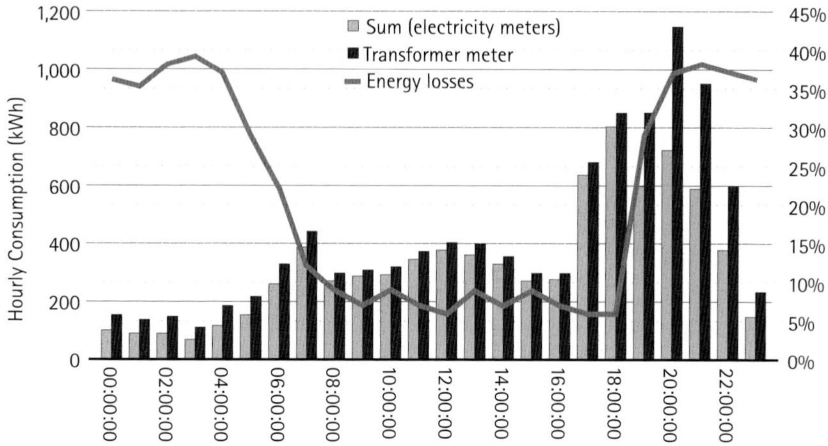


Fig. 1–31. Example of an energy balance graph

It is hard to imagine that utilities responsible for millions of meters could execute manual and individual analysis of their metering systems. Analysis systems must be deployed in order to provide frequent automatic analysis and feedback about the metering park. Important elements to be explored are load curves and other energy profiles. Intelligent and dynamic algorithms should be used, as well as automatic workflows, to ensure that all detected irregularities are treated in the field.

**Support technologies for revenue protection.** Metering circuits have to be well protected to prevent illicit activity—tampering and vandalism. Proper boxes and conduits should be deployed to avoid energy theft and to permit easy detection of tampering. Reinforced cabinets and boxes also protect metering systems against vandalism and tampering.

One of main issues associated with tampers and manipulation is the lack of legal evidence. That is why utilities are sometimes able to detect irregularities but not to recover their losses. Seals and security labels are important ways to preserve tamper evidence. They provide clear support to prove that manipulation occurred or was even attempted. It is also important to provide enough proof to distinguish between illicit access and legal actions such as mistakes made by a utility company during installation and commissioning processes.

Seal and label technologies are internationally available from basic models to high-tech ones. The utility should choose the model most adaptable to its business. For proper evaluation, it is vital that a number of features are in place. Seals and labels must be at least:

- Safety proof. For example, special attention to specifications and training should be focused on any material that could conduct electricity in order to avoid accidents.
- Utility specific and individually identifiable and traceable.
- Resistant enough to avoid false detection but able to produce legal evidence of illicit manipulation. Physical evidence is essential even if digital alarms are present.
- Usable once only and noncloneable.
- Customizable and adjustable to the environment.

An IT management system for seals is also important. For example, control is essential to know where, when, and by whom seals were installed. The lack of control and tracking methods is a common problem among energy companies and may generate huge financial losses. There is no advantage in installing seals if they are not properly managed.

Tamper detection is a fundamental problem, but there are also problems associated with costs and procedures to recover past losses. Because of this, several utilities have invested in research to develop technologies that are able to prevent tampering. Several manufacturers claim to have antitamper meters and instrument transformers. There are even manufacturers that claim they can produce meters capable of compensating for different kinds of nontechnical losses. These meters would have the capability of detecting tampers, understanding the level of measurement error produced by them, compensating for it, and keeping the meter itself within a specified metrological class.

In some situations, enforcing the security of the metering system is not enough. These are cases where energy networks should be upgraded to enable the required level of tamper protection. In electrical terms, there are several grid arrangements in the world designed specifically to avoid tampers. Some manufacturers also claim they can produce antitampering cables that are designed to prevent and detect clandestine connection attempts.

**Technical inspections**. There are basically two kinds of metering system inspections that can be executed in the field. They can be visual or technical. As the name suggests, visual inspections produce a visual overview on the metering installations and detect related irregularities. Technical inspections should be preceded by a visual inspection. These inspections consist of the execution of the technical metering tests themselves using appropriated tools and equipment (figs. 1–32 and 1–33).

It is essential that specialists are correctly trained. Safety must still be paramount for any metering action. Safety equipment, metering tests, diagnostic equipment, and tools should be provided to specialists.

The inspection criteria are often based on standards and regulations. Utilities may use these inspections to interact with customers. This would be an opportunity to provide safety analysis on the customer's metering premises, training the customers, and resolving any customer concerns.

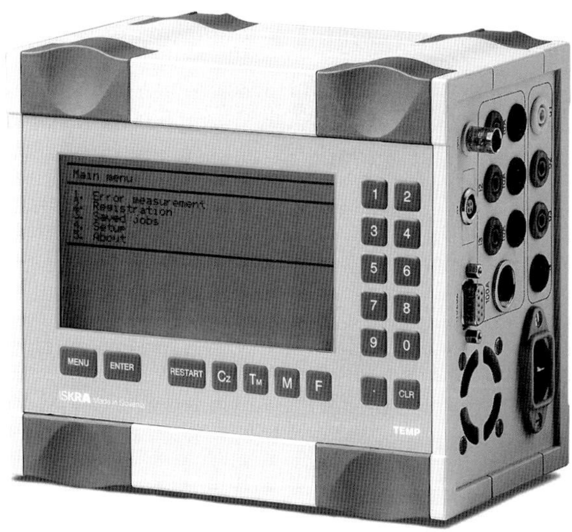


Fig. 1–32. Example of a metering test device. (Photo courtesy of Iskraemeco.)

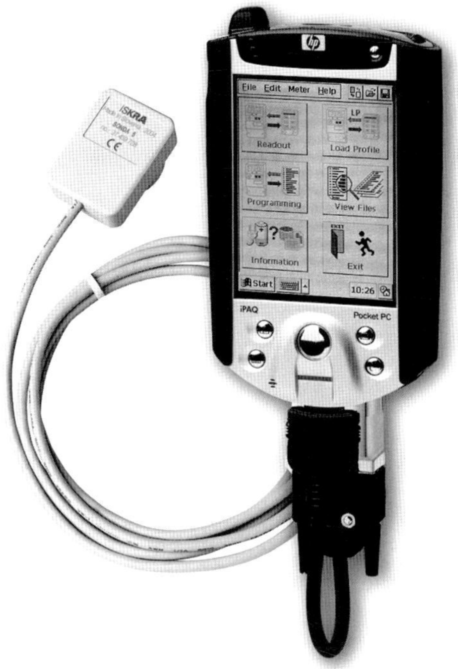


Fig. 1–33. Example of a data collection and analysis device for metering purposes. (Photo courtesy of Iskraemeco.)

# Conclusion

Metering systems have been evolving since their conception early in the 19th century. Since that time, several technologies have come into the market, thanks to research and development projects and consequent innovations.

Gas and electricity meters are available in many models and use different methods for measurement purposes. These methods change according to diverse parameters, such as the needs related to the customers' loads and utility strategies.

Meters separate the home environment from the utility environment. This boundary feature makes it an important element in providing a relationship between the utilities and their customers. Based on this core function, new payment methods have been emerging that can help customers to adapt their budgets to their energy consumption needs. They enable consumers to better understand their fuel usage by providing interaction with a number of different sales channels and also with customer display units. These technologies are quickly becoming essential.

Meters are more and more resembling computers. They are remotely managed using different techniques such as AMR, AMM, and AMI, and generate a great deal of data to be treated centrally by energy utilities. They are thus not isolated devices but part of a system: the metering system.

Metering systems have become increasingly complex and provide more opportunities to different market participants. As consequence of this, IT and metering groups must now work in better cooperation. All the utility stakeholders should also work together to enable their metering systems to become smart applications for their companies. Efficient communication and training are important tools in this endeavor.

The majority of smart metering projects in the world have as one of their objectives reducing operational losses, due to their cost to energy companies. The challenges of nontechnical and technical losses are difficult to solve, but great progress is being made with the appropriate tools and technologies.

Smart metering is not a trend but is a real solution that opens up new opportunities to utilities.

# 2

# Smart Metering Systems

In chapter 1 we discussed basic energy metering systems, their features, and auxiliary components. That information provides a solid background for understanding the concepts involved in smart metering systems presented in this and later chapters.

There is much to discuss about these systems, and there are different ways to interpret their features, functionalities, and opportunities. Practically speaking, a smart metering system is an advanced, flexible, interoperable, future-proofing system. The first three terms suggest a system that interacts with others, and this capability enables the fourth feature, future-proofing. Interoperability may also generate a practically unlimited number of service opportunities. A smart meter combines these features to provide innovative customized services and functions for different participants in the energy market such as customers, utilities, and regulators.

In another words, the benefit is all these associated systems can interact with other system infrastructures to generate new opportunities. In contrast, a single system can be limited by its predefined scope. When integrated with systems that have similar objectives or configurations, new opportunities may appear, such as cost reduction using common components or interdependent services.

Flexibility is equally important, because smart metering system implementation varies from environment to environment and is directly dependent on the needs of the parties responsible for its deployment, such as utilities and regulators. Implementation is generally customized; it is unusual to find one system implementation identical to another in the real world.

In the past, metering system components such as electricity or gas meters were seen as technical elements. This was probably due to the complexity of the early systems. Engineers built them based on technical objectives and the technology available at the time. The design and location of the electromechanical electricity meter are classic examples. Its register display format is not as customer friendly as could be designed today. Sometimes they are located outside customer premises. The question is, does this really enable customer interaction? We can easily see the answer to this question by asking another one: How many of us bother to monitor our consumption on a daily basis?

A customer display is a special part of the meter because it can provide the interaction between customers and their utility providers. The design must be adaptable to customer needs. For example, perhaps it would be easier for users to understand their consumption if it was provided as a monetary value. Meters may not be the best devices to act as customer displays, unless designed with that purpose in mind (fig. 2–1). Displays installed inside customers' homes offer opportunities for customers to better understand their energy consumption and the benefits available from new services.

Fig. 2–1. Example of electromechanical meter registers versus a modern customer display unit. (Photo courtesy of Yello Strom.)

The term *interoperability*, with respect to metering systems, generally refers to technology, but not necessarily. In order to make good use of all available metering advantages, customers as well as other market participants must interact with these devices. The customization that a smart metering system brings is an important tool to achieve this objective.

We used the meter register display and locations as simple examples, but there are several hidden opportunities. We will discuss some of them in this book. A study of these aspects will enable you to understand not only what a smart metering system is but also what it is capable of in the future.

Before adopting any solution, legal aspects must always be analyzed such as patents, intellectual property rights, regulatory requirements, and several other areas. This is valid for all the technologies, technical architectures, solutions, services, communication methods, and opportunities studied in this chapter and for any decision involving smart metering projects for energy utilities.

At this stage, let's explore the smart metering system concepts and components by analyzing the environment, related components, and functionalities. Technical architecture connects them all.

## Technical Architectures for Energy Smart Metering Systems

A technical architecture is a logical structure that provides ways for physical and logical components to interact with each other as well as with clients or providers, such as users and associated IT systems. Private and public standards are responsible for providing the rules of interaction for these components.

There are different architectures for smart metering systems, offered by solution providers and manufacturers and implemented in utilities around the world. To analyze them, it is necessary to establish a common design. Let's first look at an advanced architecture that reflects some of the main systems that may be associated with smart metering (fig. 2–2). Several layers that interact with each other in different ways form this architecture.

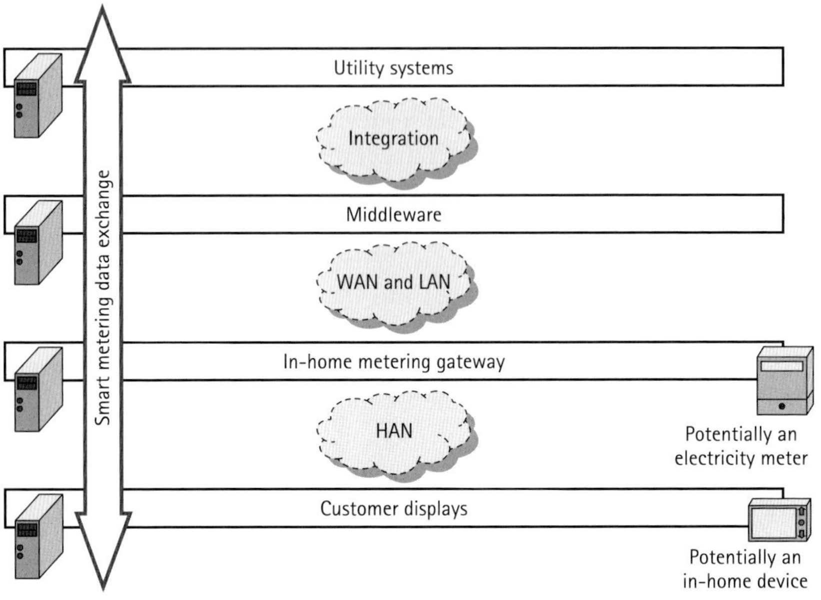


Fig. 2–2. Architectural environment diagram

These architecture layers include the following:

- **System integration:** the environment that integrates smart metering middleware with other systems. This integration is implemented with external and internal systems such as IT supervisory control and data acquisition (SCADA), billing, and customer relationship management (CRM) applications. The objective is to provide output and receive input data from and to the middleware.

- **Middleware:** a system of IT software applications and databases. Its main responsibility is to manage the smart metering devices, data, and functionalities, as well as to provide integration with other IT platforms. Middleware is often housed on servers in the corporate IT infrastructure.

- **WAN:** wide area network communication devices responsible for providing remote access to metering systems form this layer. It can be a single medium or a group of media in a hybrid architecture. WAN modems can be embedded in meters, external gateways, data concentrators, external hotspots, and other devices.

- **LAN:** generally present where smart metering architectures make use of data concentrators. In this case, the in-home metering gateways are part of the LAN (local area network) environment, using a LAN modem to obtain access to a WAN through a data concentrator.
- **In-home metering gateway:** provides interaction of the smart metering devices through different network layers such as WAN, LAN, and HAN (home area network). It is also often responsible for managing the data traffic in the WAN or LAN and HAN environments.
- **HAN:** the environment for devices installed in the customer's home such as meters, small and large home appliances, and motion sensors for in-home automation. In some implementations, HAN communication media may be the same as those used for LAN technology, but are generally different from WAN media.
- **Customer displays:** may be available for customers through different sources such as in-home display devices; existing devices in their premises such as personal computers, handheld devices, home automation panels, mobile phones, and televisions; call centers (customers could call these services for information on their meter and credit transactions); and web interfaces on Internet or intranet environments (e.g., in-home metering gateways with embedded web servers).

We discuss the individual components in this chapter. At this stage, before studying these layers in more detail, it is important to understand the way they are architecturally organized. Basic information about the above layers will help you to understand the architectural organization.

## Architectural organization

Several architectures are available for smart metering systems. Given that there is no perfect architecture, they may be adapted to different environments and needs. Architectures define the way the devices are managed and interact, and how the processing is divided in the different layers. Due to the number of possible architectures and the complexity of their system arrangements, this section focuses on just a few of the main options.

**Data management.** A middleware system is generally organized in modules. A module contains a range of similar functionalities. Examples of middleware modules are task scheduling and data security management. These modules are organized in a centralized or decentralized way.

In a centralized mode, a variety of modules and functionality are concentrated in a single middleware system. The middleware is installed on a server or cluster of servers, and all the user and system accesses are centralized. The main advantage associated with this method is data management. The reduced cost of maintenance is another valuable aspect; it is not necessary to synchronize data in middleware. A centralized database manages data collection with a global view and processing of data. It also facilitates system integration for data import and export. The data access view in middleware can give an individual data view per region even if storage and management are centralized. The downside to this method is the complexity of the middleware. The association of several modules in a single global system may add management risks.

The alternative is to decentralize management. One possible view is to organize middleware modules in a regional deployment arrangement. This method is sometimes used by utilities that serve a large area and number of customers. In this case, deployment is divided into regions, not necessarily related to political or geographic arrangements, and each region is individually managed. It provides the possibility of customized deployment per region. This strategy can be more adaptable to customers and their specific needs. Another interesting option is to deploy a hybrid option in which part of the middleware arrangements, generally the core ones, are centralized, and part are decentralized.

A different view for centralized management is the possibility of using existing applications as part of the smart metering deployment. In this scenario, some of the functions, commonly executed by middleware applications may be provided by other IT systems. This may be a very cost-effective option but can generate more complexity for data integration procedures. Examples of platforms that can be integrated are SCADA, CRM, Geographic Information System (GIS), data mining, and data security.

**Data processing.** Smart metering system processing may be implemented in different layers. The division depends on several aspects such as communications technology and IT systems. How these systems are implemented may influence what functionality is offered and the speed of global response.

***Fundamentally centralized processing.*** This arrangement does not mean all system processing is centralized, but only the majority of its core functions. The idea is to concentrate these functions in the middleware. In other words, we remotely manage the majority of meter functions. This varies from a simple on-demand task management organization, up to a full online environment. An example of an on-demand environment is when the data collector module, located in the middleware, starts the data collection procedure.

The other extreme could be a prepayment system where credit management, in special decremental tasks, is centralized in the middleware. In this case, we frequently collect register information from the metering system and calculate credit levels in the middleware. The meter then receives frequent credit updates and acts as a display only (fig. 2–3). In the event of credit purchases, we calculate all related fixed charges, fees, and other financial charges in the middleware. The credit is then periodically updated, taking into account divided amounts such as fees and fixed charges.

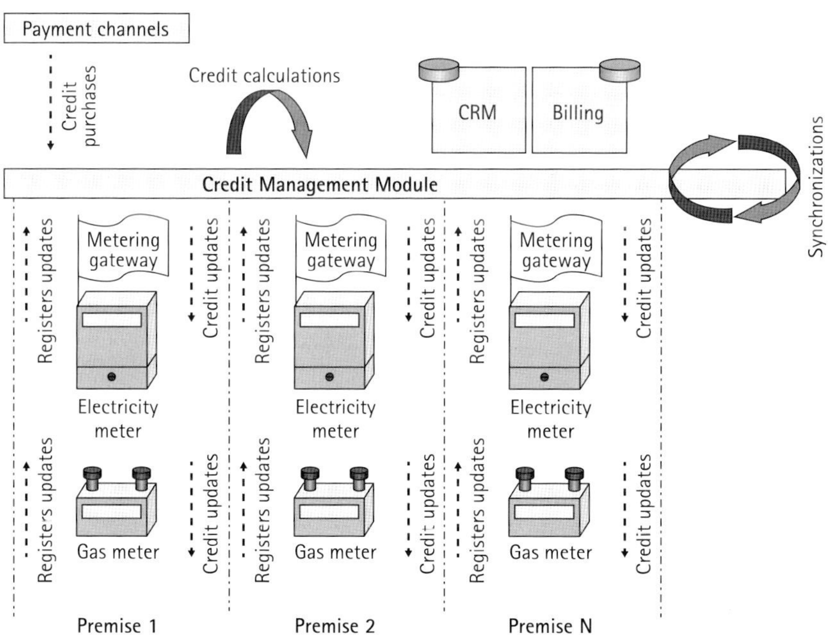


Fig. 2–3. Example of prepayment using centralized processing

An advantage of this arrangement is that it significantly reduces the level of complexity of field metering system equipment. The meters and associated devices can then have a low level of extra processing (not related to its main metrology function). We carry out advanced functions, or non-metering-related functions, in the middleware.

Disadvantages of this option are heavy reliance on centralized processing and full dependence on WAN communication media. Failures in communication may produce credit management problems, such as inefficient updates or debt generated by disconnection due to insufficient customer credit.

***Fundamentally decentralized.*** In this option, most main tasks are executed locally in the metering system devices. The meters are fully smart and execute complex calculations and functions such as monetary operations, credit management, consumption forecasts, energy efficiency functions, and carbon emission estimation.

Assuming the same prepayment example as above, all credit decrementing would be fully executed in the meter itself. The meter is also responsible for managing payment of any periodic charges like fees and fixed charges (fig. 2–4).

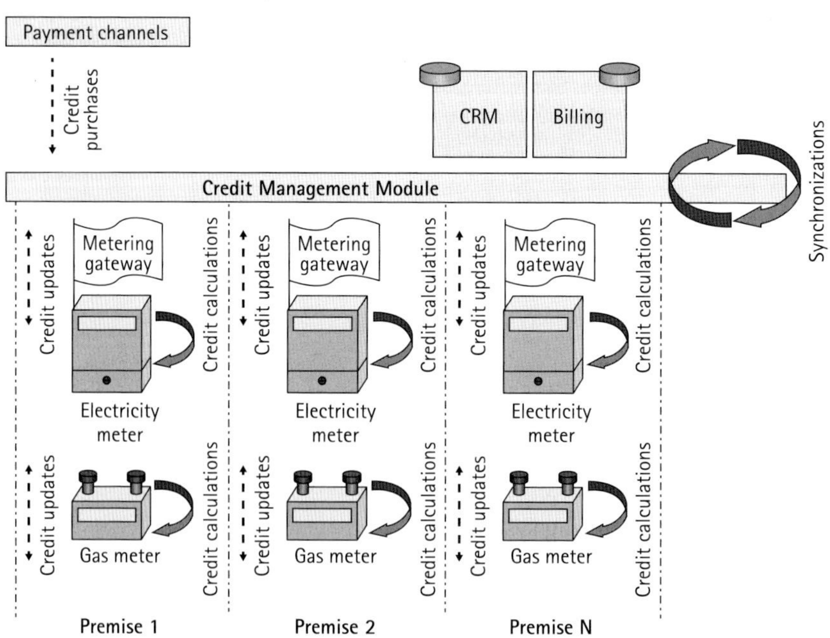


Fig. 2–4. A prepayment scenario using decentralized processing

This provides a reduction in middleware processing as the metering devices will process part of the tasks. An advantage is low dependence on the communication media. A disadvantage may be too many functions managed by the energy meters. This may result in problems in the core management functions of these devices. This risk can be reduced by implementing an embedded but fully separated management module.

***Hybrid processing.*** This combines both of the previous cases. Some of the functions are centralized and others decentralized. This configuration seems to be the most appropriate for the majority of smart metering deployments. It assumes that nonmeasurement modules implemented in the metering devices are completely separated from the measurement modules. A common implementation is an energy meter associated with an in-home gateway. This gateway is a fully separated device that interacts with the energy meter. It can be externally located or embedded in the meter.

Another example of hybrid processing would be to purchase credit locally and allow the middleware system to manage the total amount due.

**Communication Networks—Logical and Physical Arrangements.** Network topology is an important aspect of smart metering systems due to its interaction with the services a utility may offer. The topology choice may influence several aspects of the system such as communications, network robustness, and availability. A communication technology may be dependent on a specific topology.

The following sections describe some of the topologies internationally available for smart metering systems. As there is no perfect solution, the utility should carry out laboratory and field tests on technologies and related network arrangements, in order to identify the most appropriate one for its needs. Hybrid options are generally recommended to reduce associated systemic risks.

***Mesh networks.*** This is an interesting arrangement for smart metering systems. It may offer many advantages to utilities such as communication reliability, availability, and cost of modems. Mesh networks are a mature technology deployed in many industries, which is why utilities have opted to deploy smart metering systems based on this kind of network topology.

In this configuration, nodes connect to as many neighboring nodes as possible during the connection procedure (fig. 2–5). This may happen several times, such as the first time a device is connected to the network or during reconfiguration (due to node removal or constant interference).

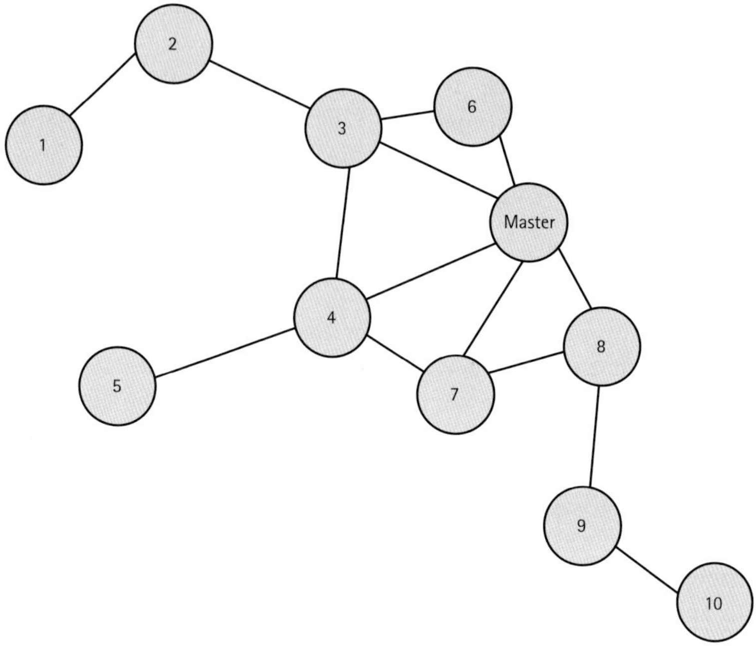


Fig. 2–5. Node arrangements in a mesh network

Data can transfer between two points by different routes at different times. This is because data are always routed the best available way. The route can change with temporary conditions on a network.

These networks can feed data concentrators or head-end devices. These support local data connection in the LAN network and route data in the WAN environment communicating with middleware. The mesh permits a node to be connected to more than one data concentrator or head-end when different routes are used. Because of these features, these networks are generally highly reliable and highly available for use.

In this arrangement, we normally need two modems per metering gateway: one for the LAN and another for the HAN. This is often due to factors such as incompatibility of the targeted media with some devices such as gas meters and appliances or because some standards limit communication inside or outside the home. In this arrangement, in-home devices such as customer display units are dependent on a meter or external gateway device to communicate with data concentrators or head-ends.

However, with a mesh environment a second modem for HAN devices may not be necessary because the same medium is often appropriate for both LAN and HAN. In this case, the data concentrators or head-ends are able to communicate with both meters and in-home devices. This can be a very cost-effective arrangement.

One of the disadvantages of these arrangements is the effective range of communication. For example, in RF (radio frequency) environments, line-of-sight signal propagation may be reduced for several reasons, such as building arrangements in urban areas, material penetration, and interference. Sometimes, this range is insufficient to form an efficient mesh arrangement. ZigBee and Z-Wave are examples of technologies that use this kind of topology.

***Point-to-point concentrated networks.*** In this network arrangement, nodes link to a central data concentrator responsible for collecting data and routing it from the LAN to the WAN environment, connecting with middleware. The range of coverage is usually restricted, between a hundred yards to a few miles. The technology is mature and is currently deployed in several countries for smart metering projects. Physically, the nodes may be connected to data concentrators using different topologies such as bus, tree, and star. PLC (power line carrier) and Wi-Fi are examples of technologies based on this concept.

Examples of bus networks are deployments based on wired networks (fig. 2–6). They are sometimes an interesting option when the nodes are close together, for example, in office or apartment buildings that group meters in special rooms or when gas, water, and electricity meters are close to each other.

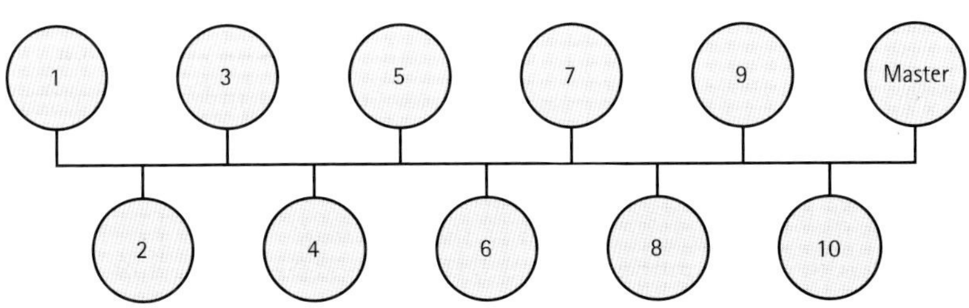


Fig. 2–6. Node arrangements in a bus network environment

An example of star applications consists of hot-spot structures that are able to provide communication to meters and other devices inside the home (fig. 2–7). In this specific arrangement, as in the previous topology, a single modem may be used for both smart meters and in-home devices. Data concentrators and head-end devices play an important role here, because thousands of these devices are necessary for large deployments of millions of meters.

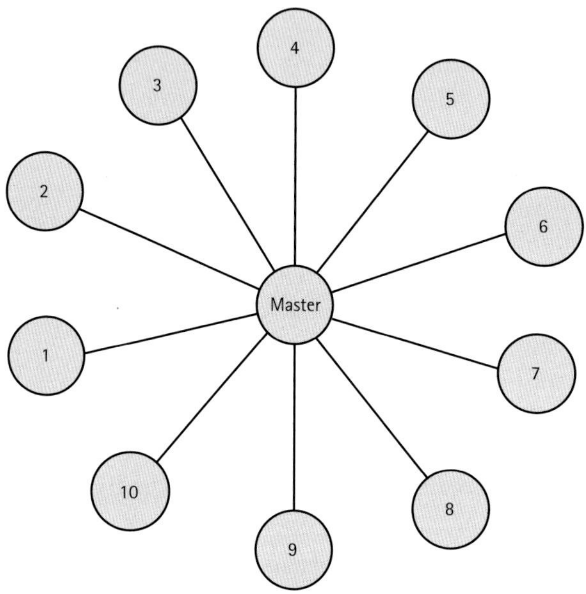


Fig. 2–7. Node arrangements in a star network environment

Among the advantages are the simplicity of use, range of technologies available around the world, and distributed processing arrangements. Disadvantages are extra maintenance points and complex management due to the inclusion of intermediary devices such as head-ends and data concentrators. Other points to note are the strong dependence of a common node, for concentration purposes, as a single point to communicate with middleware and the possible reduction of performance from use of this kind of intermediary device.

***Virtual point-to-point networks.*** In this scenario, the telecom infrastructure is invisible to energy utilities because it is usually owned, managed, and maintained by third parties. In these networks, depending on the physical topology selected, a virtual point-to-point environment is established between middleware and the devices in the field. Examples of technologies using this arrangement are SMS- and GPRS-based metering systems (SMS = short message service and GPRS = general packet radio service) (fig. 2–8). In this case, data traffic between the middleware and devices is via Global System for Mobile (GSM) communications.

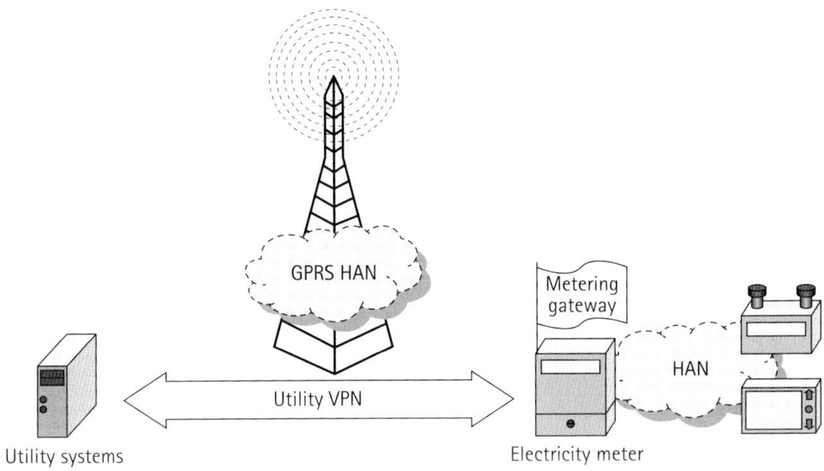


Fig. 2–8. A virtual point-to-point arrangement

Another common example uses an in-home metering gateway to provide a network address translation (NAT) service for communications between the middleware and other HAN devices such as gas meters and in-home displays. It is usually a hybrid topology because the HAN network may use several types of physical and logical topologies.

For short deployments, modem-to-modem dial-up applications are commonly used. This may be an interesting solution for trials that are not testing technology and for low-scale deployments, although it is not advised for medium and large smart metering rollouts.

An advantage of this topology is the simplicity of access from the middleware infrastructure to the devices in the field. One of the disadvantages is the dependence on a single route for communications. This problem can be mitigated by use of other media-related protocols for backup purposes with different features related to the signal strength, such as a GPRS meter that uses SMS and circuit switched data (CSD) as secondary communication options.

***Broadcast networks.*** This arrangement is similar to the point-to-point concentrated networks, but the main difference is the range of communication. In this case, the range of communication is tens of miles (fig. 2–9). This technology is commonly used for signal diffusion for television systems. Examples of implementations in the energy sector are the following:

- RF controlled systems: This is the case in the United Kingdom, where a signal is transmitted to the receiver by the transmitter broadcasting on BBC 4 long wave (198 kHz) throughout the country.
- Ripple controlled systems: This is the case in France but also in other European countries (Germany, Switzerland, Finland, UK, Greece, and others), Africa (Tunisia, South Africa), and Australia, where a low-frequency signal (175 Hz in France) is injected onto the medium voltage network and broadcast to receivers installed on the meter frame of the subscriber (or integrated into the customer's meter).

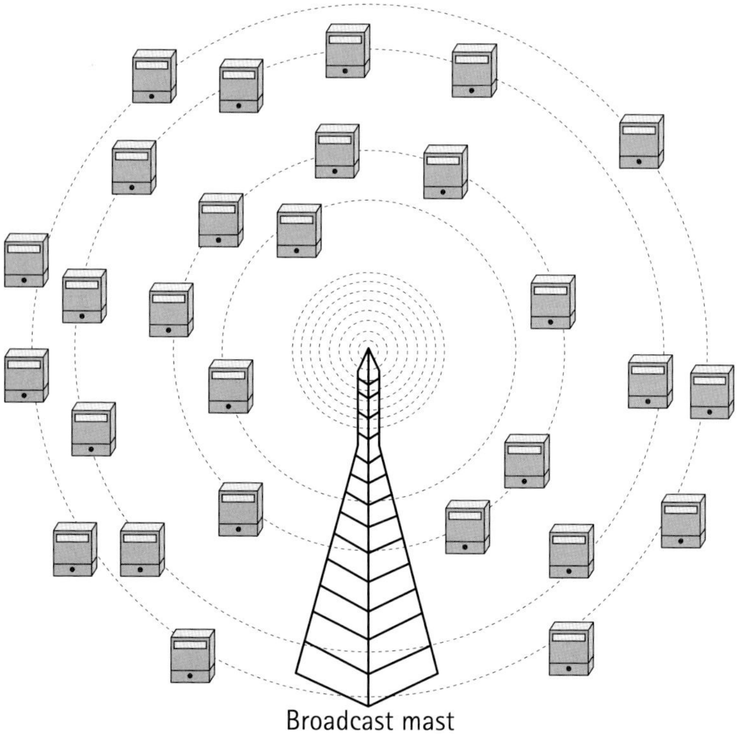


Fig. 2–9. A broadcast arrangement

Broadcast networks have been historically used by utilities as a method to provide a signal to change rates in meters or to command load management switching in the field. These networks traditionally used one-way media, but several technologies are now available with two-way capability.

An advantage of this technology is the reduction of deployment and maintenance costs, due to the low need for telecom masts to cover a utility area. A disadvantage may be low bandwidth provided by technologies that use this kind of arrangement.

## Architecture layers analysis

Now that we've covered the layers of an advanced smart metering system and their main definitions we can discuss the layers and components in more detail (fig. 2–10).

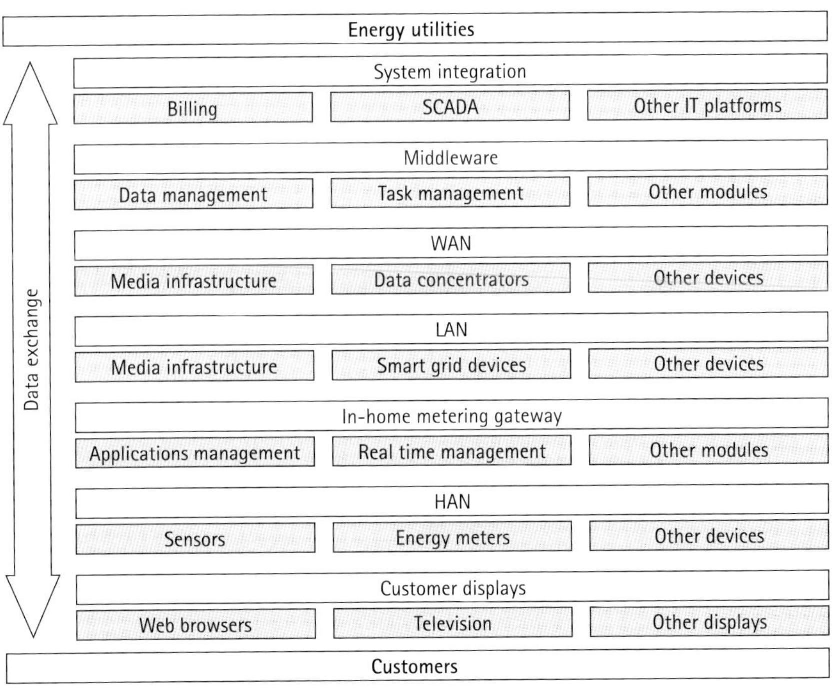


Fig. 2–10. Global architectural environment diagram

**System integration.** Interoperability with other IT systems is a fundamental feature of smart metering systems. It is generally provided via a specific integration module in middleware (fig. 2–11).This module provides options for future upgrades and other opportunities for both internal clients and other market participants. A good integration specification is also important in order to reduce implementation risks and unnecessary costs, for example by avoiding acquisition of duplicate modules that provide functions already available in the existing corporate IT systems of the utility.

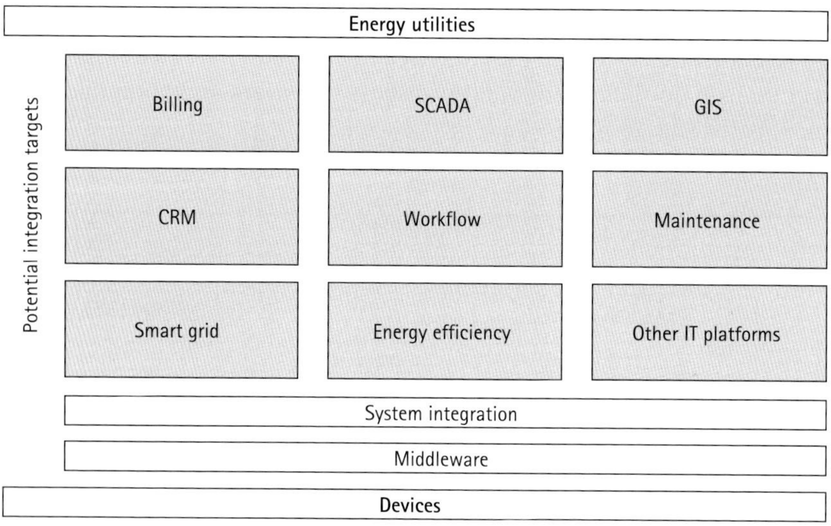


Fig. 2–11. Architectural environment diagram—system integration

It is also important to mention that data integration often provides advantages for both systems. This can be achieved in several ways such as data input, monitoring, feedback, and updates.

***Challenges and risks.*** Even with the clear importance of system integration, we can still find serious problems in systems deployed around the world. There are many reasons for this, but mainly the problems concern project specification, implementation, or acquisition of platforms that are inadequate for the utility needs. Sometimes these inadequate systems are implemented due to imprecise benchmarking exercises and consequent deployment of a project that is unable to meet the level of customization needed by the utility.

At this stage, we should reiterate that it is very rare that a smart metering system would be used by more than one utility without some degree of customization, including integration. It is important that the utility, the new integration solution provider, and the provider responsible for the existing legacy system (under replacement) must participate together in integration planning. Unfortunately, there are still cases where only the new solution provider and the utility work together to plan the procedures of integration. There are several reasons for this, such as competition among IT solution providers. This exposes the utility to integration problems and risks, such as additional setup costs or an unusable platform. In addition, more human resources may be needed to enable interoperability between systems with extra IT tools or even to manually enter the data and transfer it from one IT system to another (fig. 2–12).

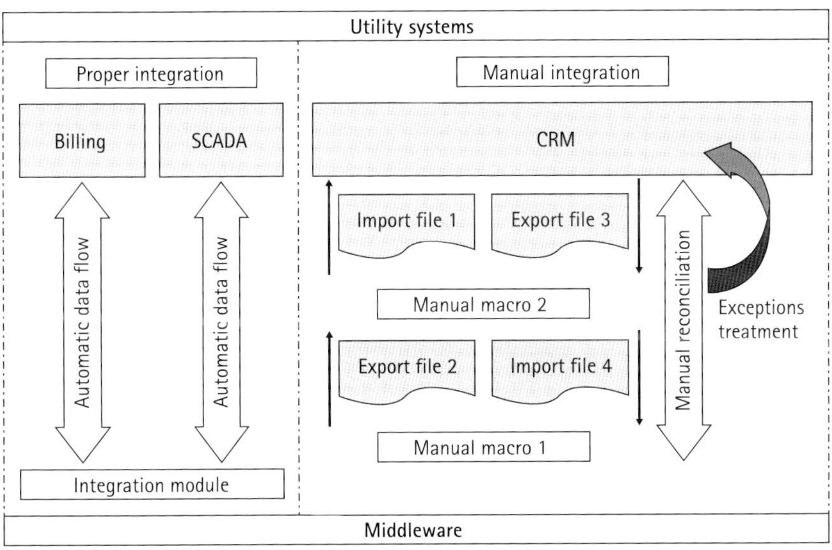


Fig. 2–12. Example of an integration problem

Integration issues can raise challenges for corporate IT departments, especially when improvised IT tools and resources are used. These tools may be designed to be temporary answers to specific integration problems, but they can become a persistent source of problems for utilities, such as breach of data security, bad data synchronization, and inability of data traceability and consequent detection of responsibility for integration problems. Even worse, when integration problems are not totally solved, they may generate the need for more and more associated data reconciliation, acquisition tools, and resources.

***Integration advice.*** It is clear that system integration may be a serious challenge to utilities. It must be well planned from project design to implementation. Efficient project management and scope definition are essential to deal with this problem.

It is also fundamental to implement a future-proofing integration module in middleware. This module should use a proper integration language standard, and provide a customized data exchange compatible with existing and future systems. These systems may be located inside or outside the corporate utility environment. The integration module should allow enough flexibility to the utility or its representative in order to avoid unnecessary extra costs for future upgrades.

Another point is the selection of a solution provider for data integration. It is advisable that selected providers have extensive experience and understanding of data integration. Meter manufacturers are not necessarily solution integrator providers; similarly, integration providers are not necessarily metering specialists. A solution for reducing this risk is to carefully define the parties responsible for smart metering implementation. Specialists in areas such as in-home automation, metering, communications, and integration may need to work together in partnership. The utility itself must not only specify but also actively participate in the implementation process. A full production environment must also be set up for tests prior to any upgrade to production mode.

It is also necessary to understand that migration from the old middleware platform responsible for managing the old metering systems to the new one will be a progressive process. While new smart meters are being installed in the field, the old meters must continue operating normally.

Integration problems must be avoided. Emergency procedures must be established in order to deal with an eventual integration crisis. Backup resources in terms of systems and people must be available and tested in the business case to avoid unnecessary risks. Utilities are likely to pay a more expensive price for failing to mitigate integration risks.

Corporate integration is a challenge, but external aspects may provide additional challenges to the utility. This is mainly due to data ownership and responsibility. An example is the definition of integration procedures involving important real-time functions such as transactions involving financial payments. In this case, there are generally at least three parties: the utility, the integration provider, and a bank. Strong traceability and acknowledgment controls should be in place, as well as efficient real-time functions such as making financial commitments. It is also important to define the responsibilities of all the integration partners during exceptions, such as temporary communications problems.

For external integrations, utilities should use an integration hub (fig. 2–13). This must be done to ensure that the required separation is made between the external and corporate environment for data security, and it also creates extra traceability between both systems. All the corporate procedures and necessary IT infrastructure for data exchange processes must be in place, such as security policing and data security devices.

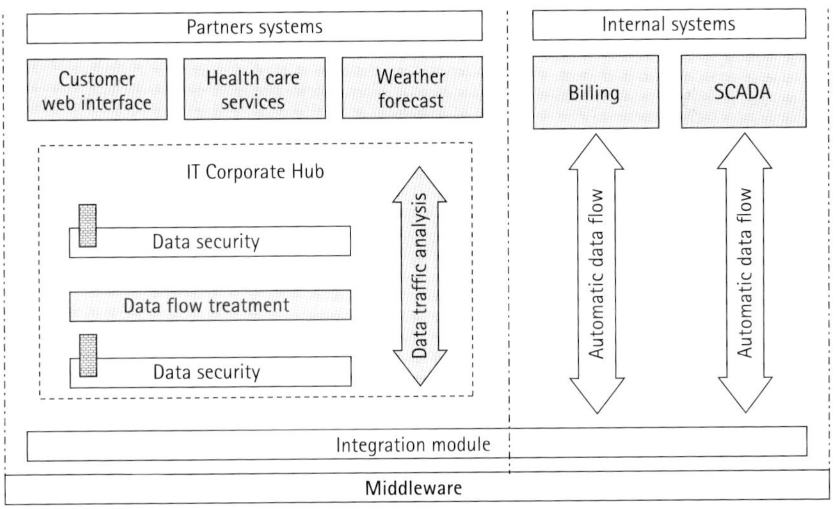


Fig. 2–13. Example of an integration platform

Although future-proofing is an essential feature of data integration modules, it is important to clearly define the scope of initial integration. External and corporate processes must then be mapped in detail. All this must be part of project conception.

***Common targeted systems.*** Several systems may be targeted for integration depending on utility needs, although there are some procedures that are always recommended. As smart metering systems generally aim to bill and provide a better relationship with utility customers, it is necessary to integrate smart metering systems with the utility CRM and billing systems. To avoid uncontrolled duplication and responsibility, data ownership must be well defined, generally in the scope of the system. For example, logical data such as rates would be owned by the billing system, and data on customer identification and behavior would be owned by CRM. This does not eliminate the need for middleware to use these data, but the system would manage it. The middleware in this case would be a simple client of these external systems and would have read rights only. It must be ready to execute commands and acknowledge orders generated by these systems, such as a remote disconnection of energy supply or delivery of a customized message to a customer.

Smart metering systems regularly provide sophisticated online services, so the related data must be manageable in an online computer environment. A clear service level agreement (SLA) must also be in place to ensure that solution providers properly perform predefined requirements. It is logical to use a supervisory control and data acquisition (SCADA) platform to manage these devices in the field and to manage the related data traffic. Among other responsibilities, these platforms are normally responsible for online device monitoring, supervising, statistical reports, data traffic problem detection, human–machine interfaces such as synoptic diagrams, event management, and other important sources of feedback to utilities.

Often, these platforms are also integrated to workflow systems that are able to automatically select exceptions to be dealt with in the field, as well as to provide feedback on respective maintenance. This integration may be provided directly via the middleware or the SCADA system, depending on the implementation strategy. Automatic workflow tools are essential to the success of smart metering implementations due to the volume of events and transactions.

Smart metering systems are also designed to improve processes within the utility company and reduce losses. However, there is a risk that existing nontechnical losses will not be corrected, or even that new sources of losses may be created during meter replacement. Bad asset management, for example with respect to the meter's location and status problems, normally increase this risk. Because of this, it is important to integrate smart metering systems with geographic information system (GIS) platforms in order to identify the point of connection to devices. They may be updated via feedback during meter replacement, for example, using a handheld unit with a Global Positioning System (GPS) module.

Several other systems may be integrated with smart metering systems such as web platforms for customers, network management systems like smart grid platforms, remote maintenance tools for IT, data mining, and systems from providers of additional services such as demand side management platforms.

It is clear that middleware plays an important role in integration. It must provide all the functions necessary to achieve its dependent goals via internal applications or by integration with other compatible systems.

**Middleware.** The middleware platform (fig. 2–14) manages smart metering devices in the field and provides the interface between them and utility applications. In another words, it lies between the devices installed in the field and management applications in the utility. The middleware platform is responsible for several tasks related to data management such as data security, event management, scheduling services, and data traffic. It manages execution of orders generated by external applications such as a billing or CRM platforms. It basically centralizes and manages any orders associated with interaction with devices in the field.

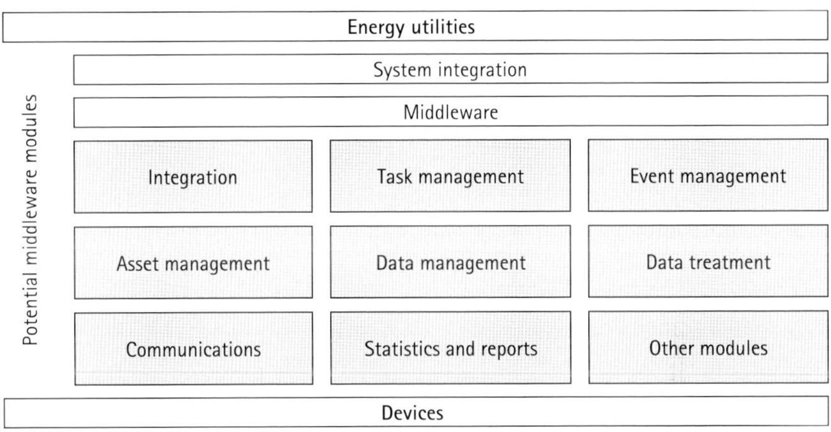


Fig. 2–14. Architectural environment diagram—middleware

Middleware is normally a modular system composed of several different management modules that interact together. The number of modules varies according to energy market needs. There are many examples of modules and functional arrangements that make up a middleware system. Depending on the solution provider and on the utility's needs, different modules and functions need to be implemented using strategies that may be adopted according to the energy market environment.

The integration management module (fig. 2–15) is generally linked to diverse agents that interact with different systems using many standards. These standards often contain detailed specifications for data transfer, such as data exchange formats and protocols. Different agents are sometimes used because of need for specific customization for interacting with diverse systems. For example, interaction with the billing system may be provided by "agent 1" using "standard 1," while the interaction with an SCADA platform may use "agent 2" given the need to use another standard for this purpose.

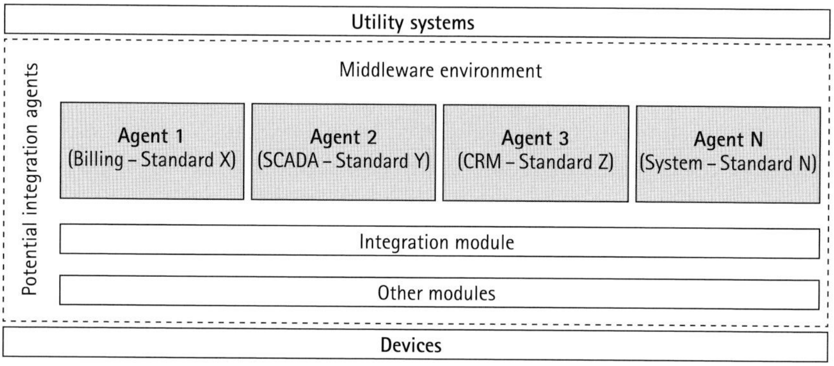


Fig. 2–15. Integration module view

Another common module is in charge of the tasks and event management. It is responsible for managing internal and external middleware tasks according to a set of predefined priority rules. These tasks may cover different orders such as data collection, device parameterization, and rate signals. Their functions may be integrated in a single module, be built in separate modules, or be part of other middleware modules, depending on the implementation strategy.

Let us assume that an order of energy supply disconnection is generated at the same time as an order for changing a rate. In this case, the middleware manages both tasks and ensures that they are treated according to their management profiles. It also monitors the task execution in its end-to-end process. It may then provide acknowledgments, task cancellations, rollback procedures, data recovery procedures, and other associated tasks.

This module is responsible for managing the events related to its environment and, interacting with the integration module, may provide feedback to systems such as SCADA or revenue protection platforms. These events may include different types of alarms such as safety, tamper detection, and diagnostics. This module also manages the event flag status and log registers.

A further module is asset management. It manages the operational aspects of its associated devices like parameterizations, firmware upgrades, and clock synchronization. It also interacts with other corporate platforms such as GIS and billing in order to obtain the necessary data to manage related assets in the field and provide respective feedback. An additional feature of this module is the ability to interact with external maintenance platforms. Asset-related diagnostics are analyzed by SCADA management platforms. This automatic process helps to ensure minimum finance or service-related losses, because problems are handled as they are detected in the field.

Data management is another function usually processed in a separate module. It is responsible for managing security, traffic, storage, traceability, and other logically related data functions. Because of data volumes, middleware generally uses a data warehouse and smart achievement tools. Assuming, for example, that energy data are collected each quarter of an hour for millions of devices, it is necessary for these data to be efficiently stored. Because of data volumes, performance tools are generally used to compress data and to archive inactive data in compact data formats.

Data processing is also an essential module. This module may be embedded in middleware or offered as an external platform. Data volumes must be analyzed, validated, and corrected if necessary in order to provide efficient feedback to utilities. This strategic feedback is fundamental to audits and decisions-making processes. Statistics, reporting, and data-mining applications are also crucial to efficiently manage these data.

What about the interaction between middleware and its managed devices? Another module manages this. The communication module (fig. 2–16) ensures future-proofing and the capability for hybrid communication with the middleware. It interacts with different WAN communication media via agents using different media standards. These agents are responsible for translating the middleware orders in a device-specific format. Thanks to this module, middleware communication may be media-agnostic, thus fully compatible with different WAN technologies. For example, middleware would be able to simultaneously communicate with devices based on GPRS or Internet broadband WAN in the field.

It is rare, or even maybe unachievable, to implement a smart metering system based on a single communication medium. This is mainly due to the different environments in a utility delivery area and media availability. This communication-agnostic capability is a key factor in the success of smart metering projects.

It is hard to forecast the future of communication technologies. Extra costs related to the replacement of middleware infrastructure as a result of media update in the field illustrate that middleware is a significant financial investment for the utility. It must be future-proof and work with the evolution of communication technologies. Utilities may pay a large price if this capability is not provided.

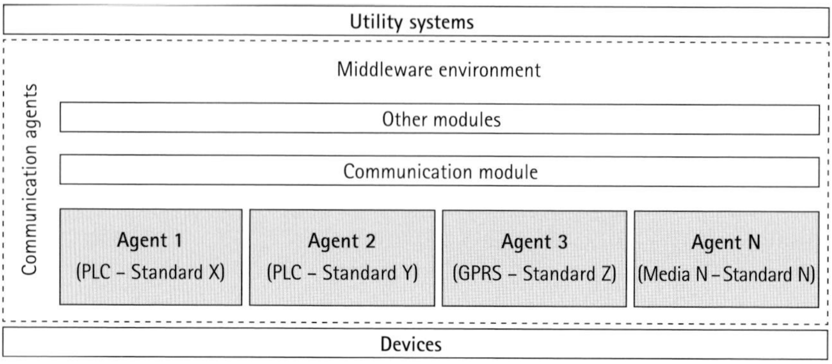


Fig. 2–16. Communications module view

In order to comply with the future-proofing requirement, middleware must be able to be upgraded with new communications agents. These modules may be built according to a predefined standard. It is better to invest in building these agents than to be forced to invest in a full middleware replacement due to obsolescence.

**WAN.** The WAN environment is an important element in the interaction between middleware and associated devices. WAN is the first layer of communication between middleware and the metering system. It ensures, depending on architectural features such as technologies and scope, direct or indirect communication with the equipment in the field. Different media may be used in this environment such as GPRS, medium voltage PLC, and Internet broadband.

In a hybrid environment, some devices may communicate directly via the WAN, and others indirectly via LAN and/or HAN. This layer is generally composed of WAN devices such as meters, in-home metering gateways, and optionally intermediary devices such as data concentrators and head-ends.

Intermediary devices are needed when the implemented technology necessitates the use of an extra LAN environment. For example, narrowband PLC technologies generally use data concentrators, and broadband PLC technologies often use head-ends to provide links for managed equipment with the WAN environment (fig. 2–17).

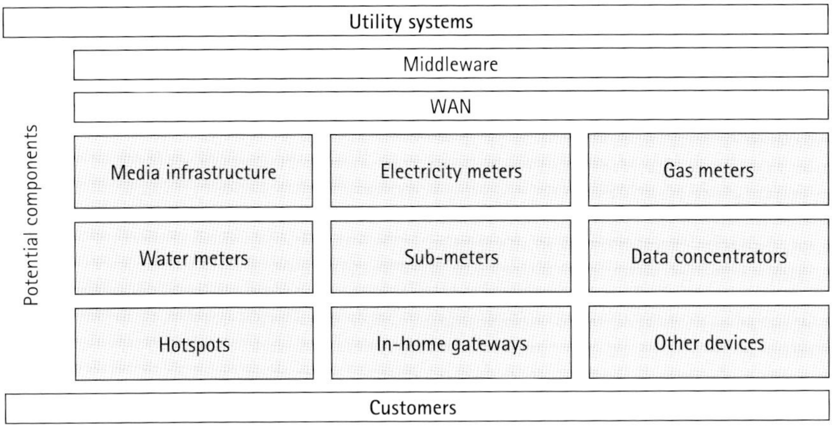


Fig. 2–17. Architectural environment diagram—WAN

WAN connections may be installed in four locations (fig. 2–18):

- Built-in WAN devices: In this scenario, the WAN device is embedded in the in-home metering gateway. An in-home gateway is able to provide WAN communication to dependent devices, such as meters and displays, without using any extra auxiliary device.

- Data concentrators: For technologies using this structure, the WAN connection is at the data concentrator. This equipment normally fully manages tasks sent to the dependent devices. It does not work as a simple router but as a data concentrator. It receives tasks, executes them, and provides feedback to the middleware. It is also responsible for pooling all dependent devices and manages its network for all levels, such as data traffic and priority controls. When associated with these devices, the in-home metering gateway only provides connection between the HAN and LAN.

- Head-end devices: Similar to data concentrators, but instead of requiring that all transactions are executed by it, a head-end device works as a communication router only. It routes data packages, providing a connection between the WAN and LAN.

- In-home modems: Homes around the world are increasingly equipped with broadband Internet access. Service providers for cable television and security alarms make use of this infrastructure to avoid unnecessary costs and complexity;

similarly, smart metering systems can also use home broadband systems. Even with the restrictions that come with these services, they may still provide valuable services to customers in some energy markets, such as device monitoring, online load profiles, and low-cost customer retention.

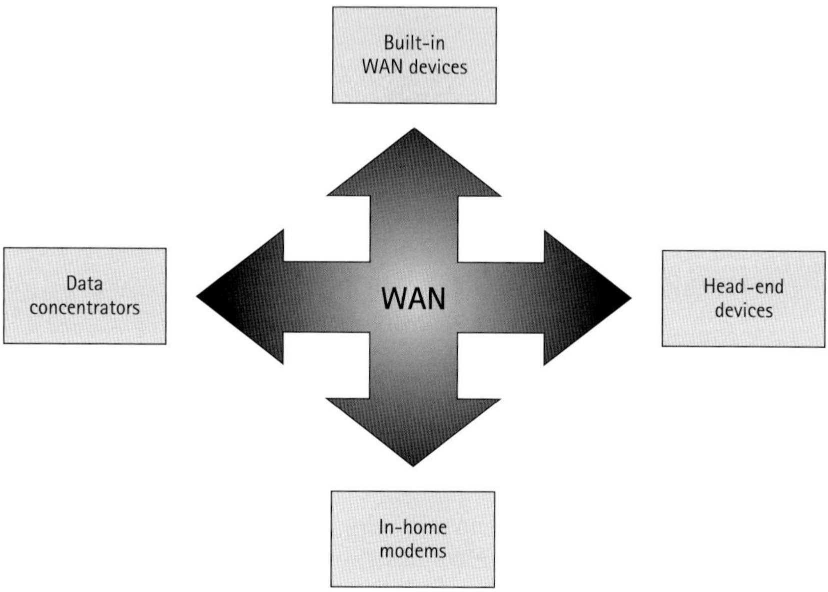


Fig. 2–18. WAN connection diagram

**LAN.** Some communication media require the use of the LAN layer, sometimes because the media are not able to provide direct communications through the WAN, or because of the cost reduction thanks to the concentration of multiple nodes on a single WAN channel. Examples of technologies using this configuration are PLC and low-power RF. Data concentrators or head-ends are necessary devices in this layer. They manage their associated LAN layer and devices as well as to provide connectivity to the WAN. They may also be responsible for advanced device management and supervisory functions such as data collection and routing, clock synchronization, and local firmware upgrade management. Local area networks are fundamental elements for this architecture.

In classical implementations using LAN (fig. 2–19), such as an implementation based on narrowband PLC for electricity meters only, the LAN devices are the last point of communication. In this case data concentrators centralize all advanced tasks for management of local nodes, although other advanced scenarios are possible.

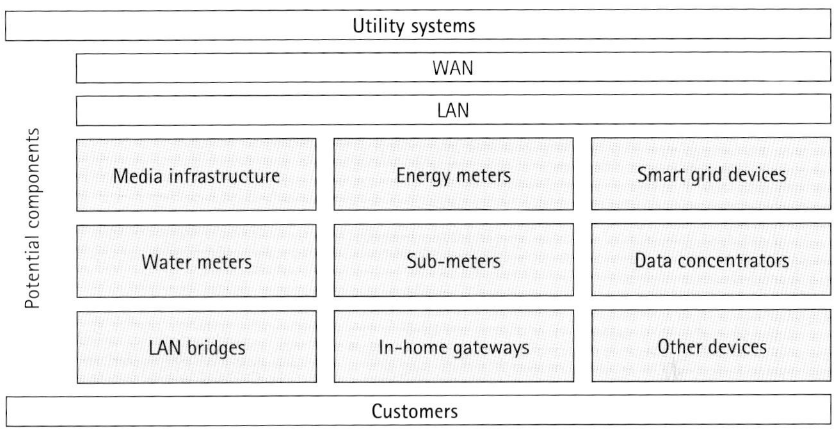


Fig. 2–19. Architectural environment diagram—LAN

Assuming that gas meters are added to the above scenario, electricity meters, acting before as simple connection nodes, would now also become gateways (fig. 2–20). Therefore, they would be embedded with additional in-home metering gateway functionality. They would then be able to provide data management for both electricity and gas meters, and would be managed by data concentrators. The local intelligence would be shared between both data concentrators and in-home metering gateways.

These electricity meters would therefore be responsible for managing their HAN layer and connecting their devices to the LAN. Of course, they would still be responsible for their own function of measuring electricity. The data concentrator would then be the final link with the WAN. In this example, data sent by the middleware could be then transferred through different devices in the WAN, LAN, and HAN to reach its final destination.

With head-ends, for example, in a broadband PLC implementation, the previous example can be managed even more easily. Given that these head-end devices generally act as routers in applications for electricity meters only, the intelligence is generally shared between the middleware and the meter itself. When gas meters are added to these scenarios, the electricity meter becomes an in-home metering gateway, which then centralizes the local management functions and associated intelligence. In this case, the head-end device would continue to act as a router, providing connection from the in-home metering gateways to the WAN.

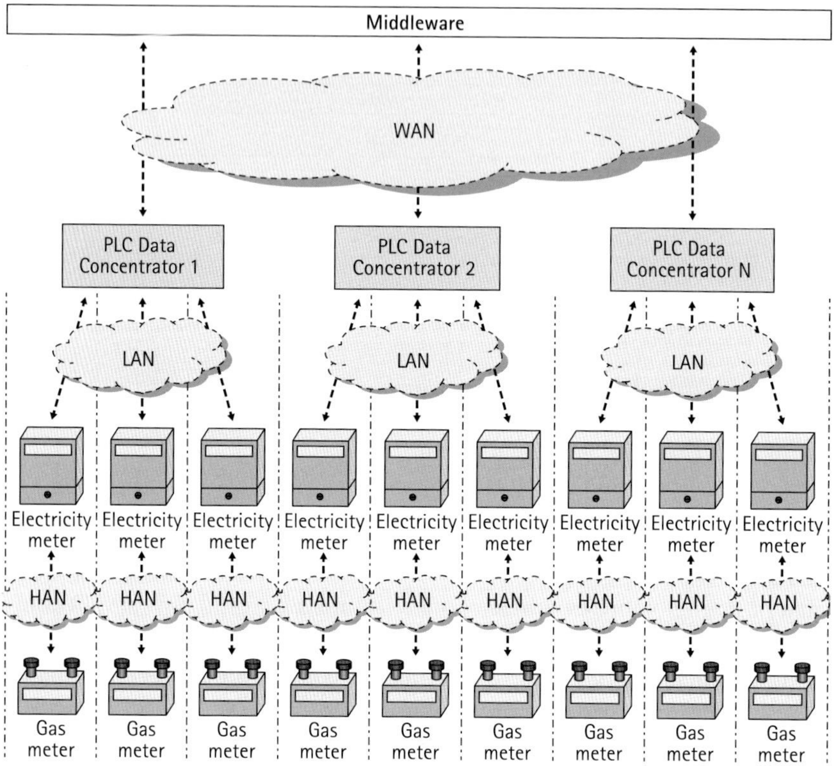


Fig. 2–20. Architecture using data concentrators and in-home metering gateways

***Bridge devices and their role in integrating different networks.*** Bridge devices are used in smart metering LAN layers. They are often used to provide interoperability between two different networks and reduce the level of complexity of the in-home metering gateway. Here, nodes of two different physical networks may be logically managed by the same infrastructure.

To illustrate this, consider the situation where a utility has deployed an architecture using narrowband PLC for their electricity meters only, without using in-home metering gateways. Therefore, the electricity meters are embedded with a PLC modem only. Now consider that there is an opportunity to manage extra gas meters from another gas utility. There would be a significant impact if the gas utility decided that an extra medium was necessary to provide communications to their gas meters, such as low-power RF. The utility would have to retrofit its existing meters with in-home metering gateways or replace them with a new model with embedded features. This could ruin the utility's business model.

A solution for this could be to install bridge devices (fig. 2–21). They could be strategically installed at different points on the customer premises by the gas utility during the replacement of their gas meters without any need for direct intervention on the electricity meters. These devices would provide the liaison between the PLC network and the low-power RF network. In this case, they would be PLC-RF devices.

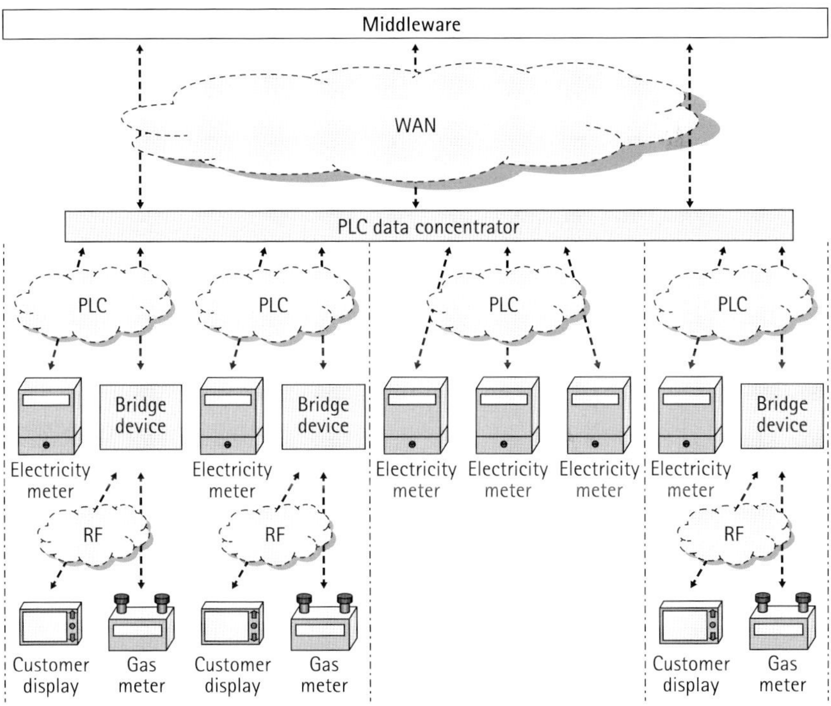


Fig. 2–21. Architecture combining bridge devices and data concentrators

Assuming that they have spare processing capability, the data concentrators could also be used to directly manage the gas meters. There could be an opportunity to use these bridge devices to connect data concentrators to other devices in the home, such as customer display units and smart plugs. In this case, the data concentrator would be a wide-scale multihome metering gateway. This global gateway function might be inappropriate for some applications using many online or real-time services but could be appropriate for several others. This would normally yield cost savings due to the low complexity and cost of these bridge devices. This would also provide for future-proofing of existing single-fuel architectures without in-home metering gateways.

***Reasons to deploy a LAN infrastructure.*** Up to this point, you may perceive this extra LAN layer as extra complexity in architectural terms. It clearly generates some complexity for utilities, although other technologies, such as WAN only, also produce complexity. The main concern is not the telecom infrastructure, because it will always be needed, but other aspects such as the following:

- Should energy utilities deploy a new infrastructure or use the existing infrastructure offered by telecom providers?
- If an energy utility decides to deploy this infrastructure, should utilities manage it or leave it to telecom experts?
- Should the utility share this infrastructure or use it exclusively?
- If the utility decides to share this new infrastructure with others, what are the associated opportunities?

These questions depend on the strategy of the utility and market arrangements. Smart metering project teams should answer these questions using tools such as business cases, benchmarking exercises, and technical reports. The following chapters of this book provide some of the key risks and opportunities for smart metering projects and project management advice that may assist you and your team.

***LAN and the smart grid.*** In the electricity domain, the smart grid is a key point. Smart grids are now being implemented around the world. They focus on, among other aspects, the use of grid sensors, switch devices, and automation and optimization of grid operations for energy utilities. This subject is discussed in more detail later in the book, but for now, it is important to understand that smart grid implementation strongly depends on the choice of technical architecture.

Smart grid devices may benefit from using a smart metering infrastructure, providing a more cost-effective deployment. The decision to use the LAN layer could be very important, for example, in implementations of smart metering systems based on low-power RF mesh architectures. In this case, smart metering and smart grid devices could use the same infrastructure (fig. 2–22). In terms of total costs for the utilities, the impact of reusing the telecom infrastructure might be significant.

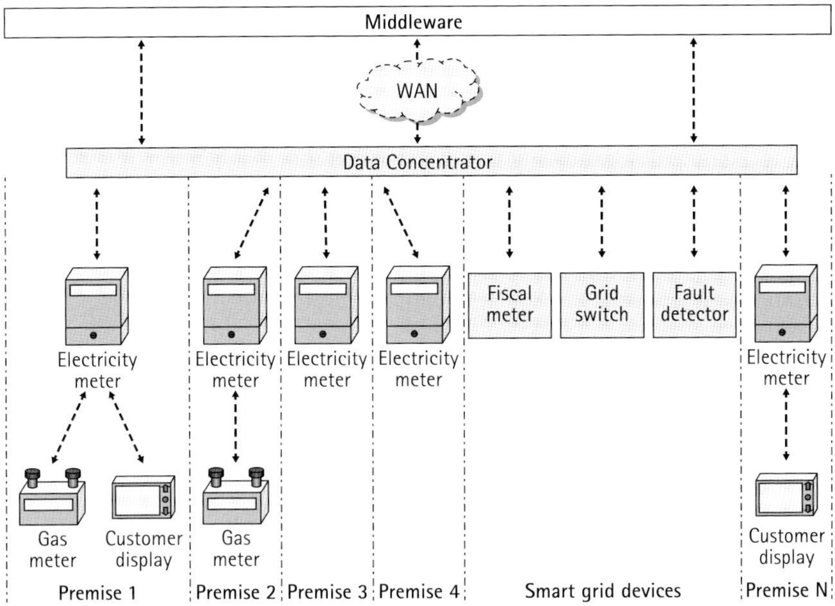


Fig. 2–22. A smart grid scenario

It seems hard to imagine the implementation of smart grid projects for energy utilities without the use of LAN networks. This is because energy networks not only have centralized points such as electricity substations but also contain a vast number of isolated points on the grid such as boundary energy points, automatic grid switches, transformers, and grid sensors. The LAN decision may be a key strategic factor for energy utilities.

It is clear that smart metering projects are not isolated from other utility or energy market projects. They may generate impacts on several other segments and initiatives. Infrastructure decisions may create additional complexity in regard to deployment and utility management, but also opportunities.

Because of the costs associated with changes to the physical architecture, it is important to the utility to carry out robust future-proofing evaluations of new strategic services and opportunities before making the decisions about the most appropriate architectural design. It is better to choose correctly than to regret it in the future.

## In-home metering gateway

***Electricity meter evolution and new concepts.*** The next layer to examine is the in-home metering gateway, but before discussing this gateway, we need to understand why it has been recognized as essential. This will help us understand the concept behind it and its opportunities.

For many years, it was very common in metering systems to use the electricity meter simply as measurement equipment with a communication device. As a communicating device, it could be remotely set and provide data to an IT application. This pull scenario may work for some devices with simple configurations, but it is unlikely to work for complex scenarios involving millions of online devices. This also created a dependence on utilities with proprietary solutions and sometimes caused the premature replacement of metering technologies.

The first initiatives to create a more intelligent and manufacturer-independent device were with meters used for measuring industrial and other high-energy-demand customers. There were fewer of these meters, but they were more expensive and had more features.

This market competition generated new requirements and opportunities for utilities as a result of competitor initiatives, increased customer needs, and regulatory goals. Energy utility decentralization programs were also introduced and, with this, branch independence, such as energy, business, and network branches. Because the branches were fully independent entities, different needs per branch also emerged in the energy utilities. The utilities needed to provide new services to their customers, because even though they were fewer in number, they represented a big share of their energy bill. This level of consumption per customer helped the utilities to invest in research and development to build robust meter platforms that were able to offer more innovative features.

An example of meters built for this market is currently in use by EDF in France to measure customers with contracted power above 250 kVA (fig. 2–23). This meter is known as Interface Clientèle Emeraude (ICE). Thanks to its flexible operating system and to a customized interface developed by EDF, the ICE meter is able to manage rate-based applications. This software interface is known as PRISME, and thanks to it, these devices are interoperable and future-proofed, and application updates can be remotely downloaded to them, such as for rate and display management. This concept differentiates the hardware base of the meter from its functions provided via its fully programmable software environment. Also, the meter is able to manage advanced functions such as:

- Harmonic analysis
- Power quality monitoring
- Load profiling

- Open protocol
- Advanced measurements and metrology
- Backup management

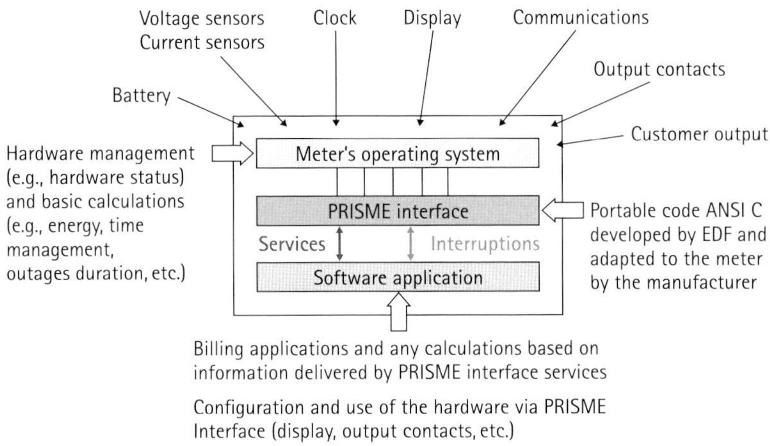


Fig. 2–23. ICE meter concept in use by EDF in France

Among other features of this type of meter, one of the most important is the capability to fully separate measurement features from the nonmetrological functions. They could then be certified by the governing authority but allow some freedom for other nonmetrological services to be offered, giving supplier differentiation in the energy market.

Meters with embedded interoperable operating systems began to look more like computers. This innovative concept spawned two important divisions: one between the metrological firmware and the applications environment, and the second between the meter and the metering gateway. The metering gateway became a separate component with fully separated functions but located in a very strategic device, the electricity meter. If deployed in the residential market, these gateways and their respective applications could be used for different purposes such as managing appliances, in-home automation, energy efficiency services, and product rates.

The electricity meter is possibly the most strategic in-home device available worldwide. They are present in every location a utility company provides service. They are nearly always supplied by alternating current

(AC) and operate through the direction of the utility company, completely independently of customers or end-users. This independency allows the utility to temporarily disconnect service due to nonpayment or for other purposes. Different service providers around the world invest billions every year to install equipment in homes, but electricity meters are always there, and by using the right technology could provide additional services in the home.

***Understanding the device and its different layers.*** In-home metering gateways are devices that are customized for complex in-home architectures and are able to offer necessary interoperability with other in-home applications. This integration feature is a key factor for the longevity of these systems in the field, and it negates boundary or scope limitations as technology and market needs evolve over time.

The in-home metering gateway is one of the most important devices in a smart metering system and it is generally embedded in or fully connected to electricity meters. It is responsible for managing the HAN layer and provides its associated devices with connection to the WAN. In-home metering gateways may also be the connection between the LAN and HAN environments for some architectures that depend on LAN media, such as low-power RF.

An in-home metering gateway has very similar functions (fig. 2–24) to those of middleware, although its functions are proportional to its level of responsibility. While middleware may be responsible for millions of devices installed for millions of customers, it is unlikely that an in-home metering gateway would be responsible for more than tens of devices for a single customer. This device is a key component of the in-home environment and ensures the future-proofing of smart metering architectures.

Assuming the existence of a high level of processing and intelligence in middleware, why share this in the home? There are several reasons: among them, speed in managing in-home services such as alarm systems. Another encompasses individual customer needs in competitive markets. Customers are naturally different. Therefore, the way one group of consumers reacts to energy-saving programs may be very different from how others react. Understanding these differences is challenging for utilities, but they are key points for customer retention. In addition, who better than the customers to understand their own needs? The in-home metering gateway allows maximum flexibility of customer-targeted services.

This device is embedded in an operating system and the utility is able to download applications to it (fig. 2–25). This capability of managing applications is essential for flexibility, integration, independence, and future-proofing.

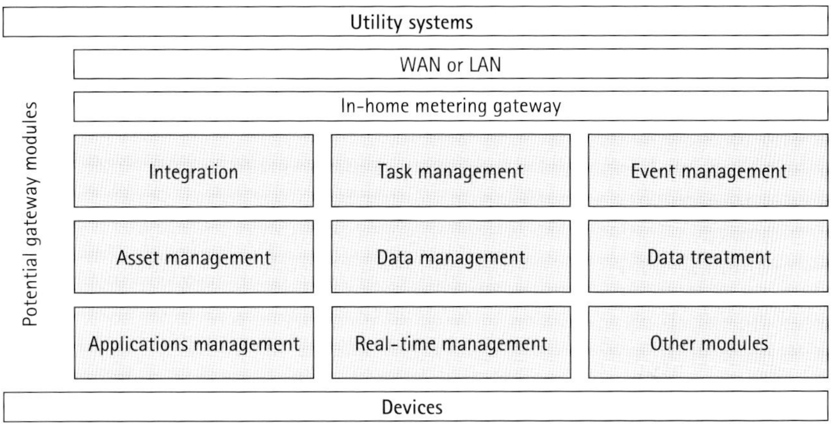


Fig. 2–24. Architectural environment diagram—in-home metering gateway

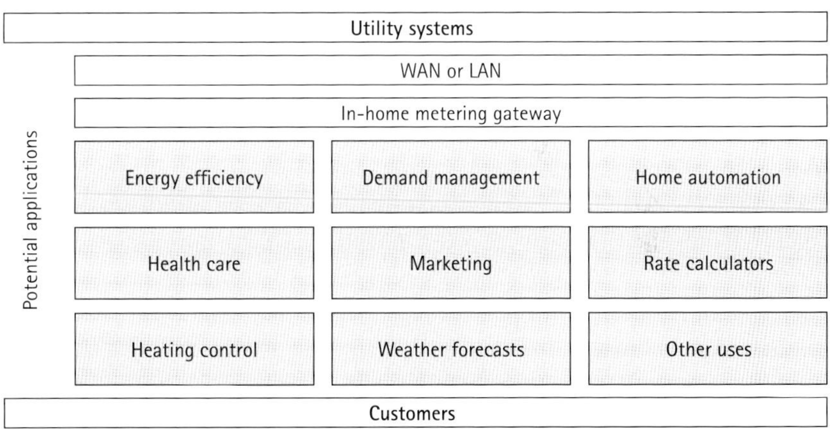


Fig. 2–25. Potential applications for use in an in-home metering gateway

Another point, slightly different from middleware functionality, is the capability to manage real-time services. Safety and health are paramount. They cannot wait! Given that all energy-related services require safety procedures, software applications in the in-home metering gateway must be able to count on real-time support of their service tasks. Transactions must also be the most user-friendly and intuitive possible; task speed is sometimes key in this process.

As in middleware, data processing plays an important role. Intelligent validation algorithms can be built to diagnose problems and enable automatic solutions when possible. Remote intervention is also vital. This equipment is basically a computer, and computers in a network must be managed remotely by specific maintenance protocols, IT tools, and help desk specialists. Due to the associated costs, it is essential for utilities to minimize field visits. Remote IT maintenance, device self-diagnosis, and device self-correction are indispensable features. This is a new era for utility companies, requiring new corporate organization and vision if they want to survive in competitive energy markets.

Computer applications sometimes crash due to exceptions. In-home metering gateways are also susceptible to crashes, and every precaution must be taken to avoid them. If this happens, it can cause several problems, such as extra cost from site visits and a low level of service quality. Watchdog events and intelligent correction algorithms must be implemented. Computational crash events must not happen!

Several devices in a home can be integrated with these gateways. For example, televisions and local computers may be used as in-home display units. The majority of customers use them daily. Heating and cooling systems can be controlled using rates and carbon emission targets. Appliances such as washing machines may also be controlled. Avoidable standby loads can be reduced for several other devices via automated processes.

It is clear that we need an efficient integration module and interoperability (fig. 2–26) approaches for current and future devices. For future devices, agent arrangements are essential, similar to the role middleware plays. Agent arrangements provide customization of in-home devices for integration. While one agent might provide standards for integration with a television, another could provide this integration with a computer via a security dongle device.

It is important that this device offers secure integration ports for future use so that it can be locally upgraded with new modules such as communication modems or memory sticks. This is needed because of the evolution of technologies and needs. The communications module (fig. 2–26) is therefore essential because it allows interaction with new technologies. It also uses agent arrangements such as those in middleware. While one module provides the communication standards for a Wi-Fi device, another can provide it to an in-home ZigBee device.

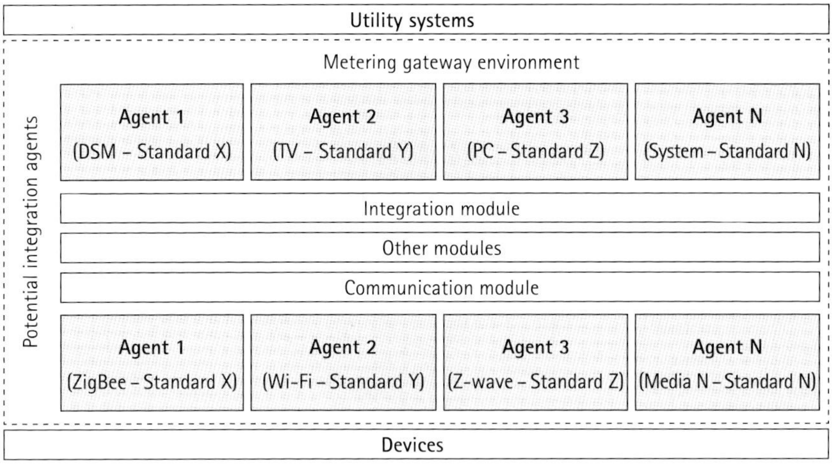


Fig. 2–26. Example of integration and communication module

Task management, event management, and data management are other modules. The volume of in-home transactions is very high as a result of the level of updates and transactions. For example, in-home displays, connected to energy meters via the HAN, are generally updated every few seconds. These management modules must provide online analysis and reports to the smart metering system.

Asset management has additional functions in relation to middleware. In the HAN several devices must somehow join the network. Because of complexity and risks such as data security, this joining process requires close attention. The module must have appropriate allowance levels for compatible and permitted devices. Disallowed devices will then be rejected. Local assets must also be locally managed by the in-home metering gateway, in addition to management already provided by middleware.

Given that middleware and the in-home gateway sometimes play similar roles, it is important to allocate appropriate bounds and responsibilities. They must operate in partnership and be complementary, with division of responsibility for individual tasks. Unnecessary duplication of tasks must be avoided.

***Installation strategy.*** An in-home metering gateway may be located in many places (fig. 2–27), although some are not recommended for some services a utility would like to offer its clients.

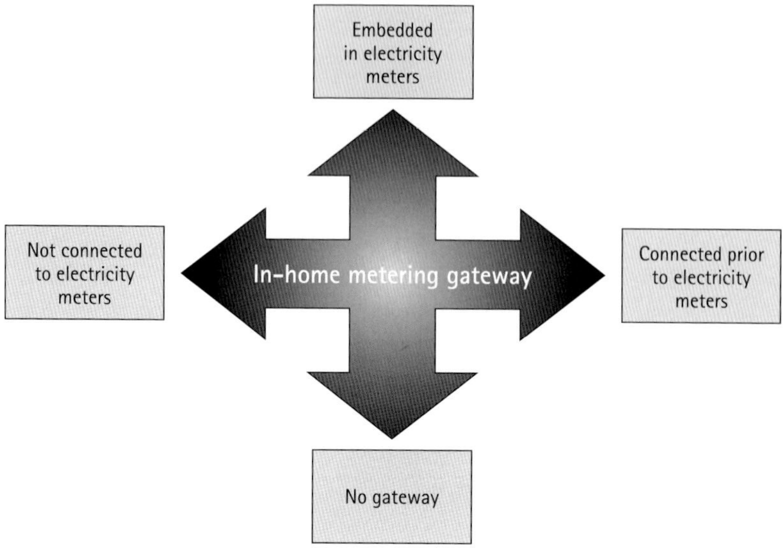


Fig. 2–27. Potential locations for in-home metering gateway

For example, installing this kind of device in a gas meter is not recommended because a gas meter is usually powered by its DC battery. There are also safety-related risks not yet solved.

Another possible location for an in-home metering gateway is inside the home (fig. 2–28), separate from the electricity meter but connected to it on the load side. This is used in some countries for specific deployments restricted to services that do not involve electricity switching services. This restriction is due to problems caused by the reconnection of electricity supply. For example, in the event of a disconnection generated by middleware or by the meter itself, the gateway would be without AC supply. Even when a battery is used as backup, there is a cost.

Based on this, the best installation for this device seems to be embedded in the electricity meter or connected to it from the power supply side. When embedded, it is important that both metering and gateway functions are provided by two fully logical and physical separated devices, even if sharing the same box. If fitted in the same box, there may be reduction of costs; for example, there is no need for an extra compartment for the gateway device. It is also important to ensure safe, robust, and antitampering connections between both devices. A single in-home metering gateway per premise is recommended (fig. 2–29), even for dual-fuel customers. This can reduce the risk of local interoperability issues, unnecessary duplicated controls, and the extra costs of using more than one device for the same function.

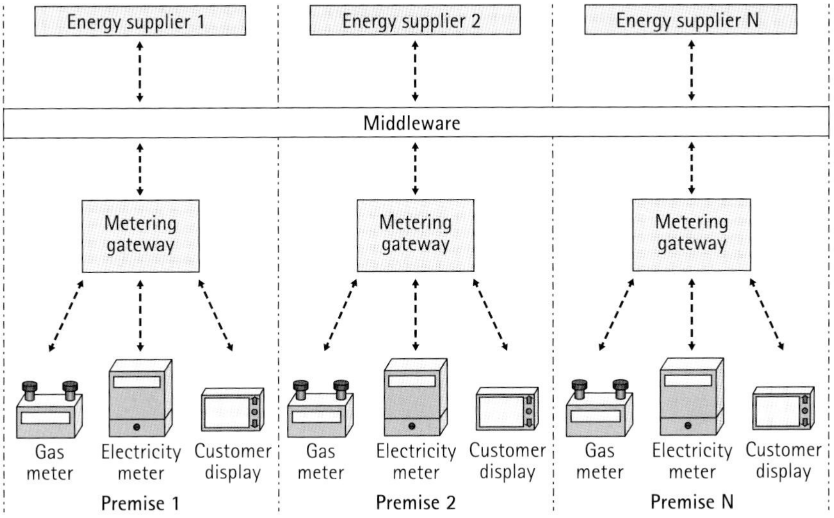


Fig. 2-28. In-home metering gateway separated from meters

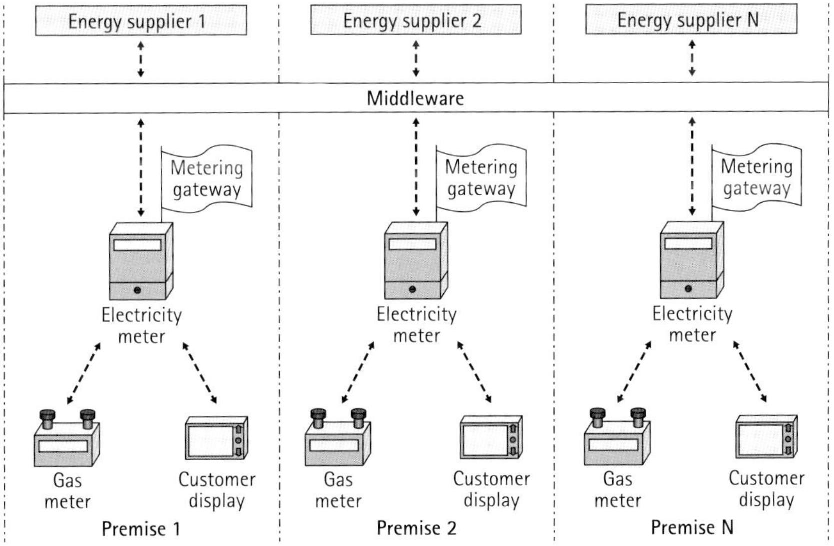


Fig. 2-29. Example of in-home architecture environment with single in-home metering gateway

Even though a good option, an in-home metering gateway may be a problem for gas suppliers due to its dependence on the electricity meter. An example is an energy supplier owning only the gas meter in a dual-fuel premise. In this case, that supplier becomes dependent on a competitor allowing connection through its electricity meter.

Another potential architecture may use more than one metering gateway provided by different suppliers in a home (fig. 2–30). It is then necessary to ensure minimal interoperability between both systems. This is important, for example, to avoid extra costs with change of supplier, or to provide a single energy display to customers.

Technically speaking, in-home metering gateways are not mandatory for all smart metering architectures, although the absence of this component puts at risk the future-proofing of the system in relation to new services, opportunities, interoperability, IT upgrades, integration, and several other aspects. Thus they are highly recommended components of a smart metering system.

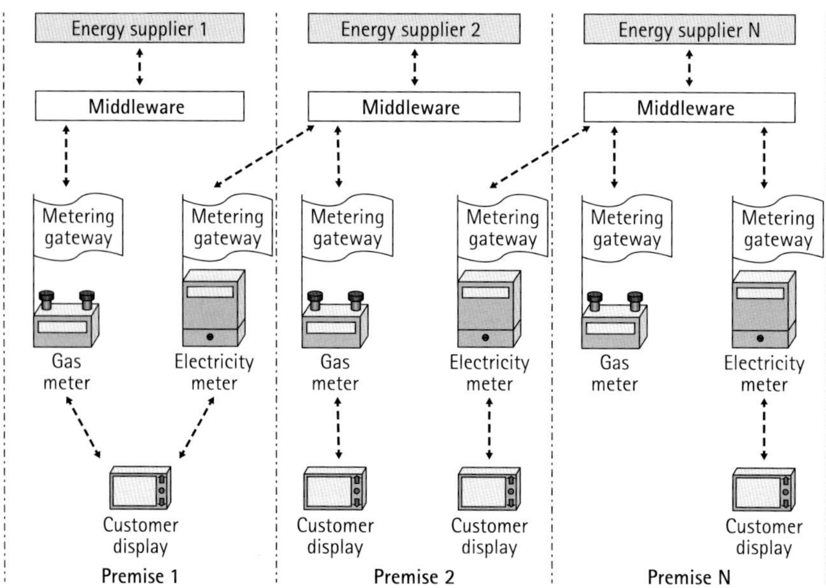


Fig. 2–30. In-home architecture environment with multiple in-home metering gateways

**HAN.** This layer (fig. 2–31) comprises the HAN infrastructure and all devices managed by smart metering systems such as energy meters, other meter types such as water meters and submeters, handheld units, sensors, home automation devices, appliances, and customer display units. Interoperability becomes a primary feature, with clear need for integration.

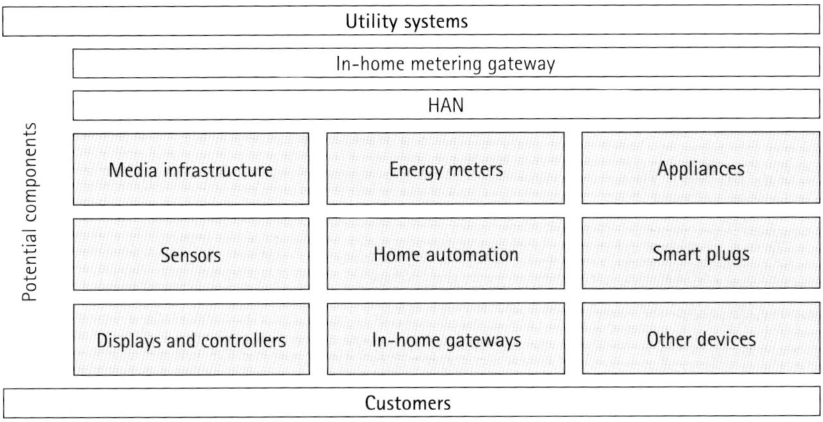


Fig. 2-31. Architectural environment diagram: HAN

A single medium does not necessarily provide the network infrastructure part of this layer. On the contrary, it tends to be composed of a hybrid combination of media. This is due to the difficulty of forecasting communication technologies to be used by different market segments that may be integrated such as appliances, meters, and in-home automation. It can be expected that an in-home metering gateway provides interoperability using different HAN media. The media devices embedded in it do not need to be part of its initial deployment. Depending on the business case, this capability may be upgraded for specific targeted customers that the utility would like to offer a specific group of advanced services. For example, assuming a HAN using one medium was massively deployed to all utility customers, they could be offered a package of services. The quantity of services could be then limited by the interoperability this medium is able to provide. Available compatible devices would define this boundary. When it becomes necessary to provide a new package of services to a market segment or niche customers, these targeted devices could be physically and logically upgraded with a new HAN using a second, different medium. New agents and applications could then be downloaded to enable interoperability with the new scenario. This kind of segmented option is sometimes important to business cases and to the customization of solutions according to individual customer needs.

Sensors are important devices in this layer, too. They can be used in the home to provide feedback to smart metering systems. Interoperability is also essential for these interactions because sensors are normally used by several systems simultaneously. This is important in terms of project costs but also to reduce complexity of the system by avoiding unnecessary duplication in the home. For illustrative purposes, assume temperature sensors are installed inside the home in order to provide temperature feedback per home environment for a heating system. Assuming they are interoperable with the smart metering system, we can avoid duplicating these devices if the utility decides to offer an automatic heating controller service via its smart metering system. In this case, these sensors could also provide the new in-home controller with temperature feedback. This controller and its respective valve devices could then be managed according to environmental heating needs, the best rate options, and defined load distribution criterion. The controller could also interact with the existing customer display unit, thus avoiding installation of a new device.

Processes can be fully automated. Clearly, machines are better at performing repetitive tasks than humans. An example is energy-saving devices. A smart plug application should have the ability to work based on predefined profiles. Assuming, for instance, that a customer usually wakes up every working day at 6:30 a.m. and goes to bed at 10 p.m., the energy-efficiency application running in the in-home metering gateway should be programmable according to this profile. In this case, the standby loads connected to these devices could be eliminated, at least for this period. Assuming that these devices have a basic measurement capability, they may automatically turn off unnecessary standby loads. In the event of an exception, the customer may opt, via simple keypresses on the display unit, to keep these devices in operation for a while. This is the kind of process that requires automation because it is unlikely customers would remember to disconnect all the unnecessary standby loads every night and reconnect them every morning.

The media used in in-home devices should be evaluated for self-consumption. For example, some chips enable internal energy counters to analyze consumption generated for communications.

Several other in-home devices may also interact in the home, but at this stage it is important to understand the fundamental needs of the HAN layer to provide required levels of interoperability, speed, and flexibility to support the in-home applications and devices.

***Handheld units.*** Providing site visits to customers' homes is something all utilities would like to avoid because of the associated costs and the inconvenience to customers. When field maintenance or installations of new smart meter devices are required, it is essential to make site visits with the lowest possible impact for all the market participants, both customers and utilities. Several aspects can be optimized such as installation time, data recovery, fault correction, efficient commissioning, meter location, and remote access in the field. All these points may be directly dependent on efficient field service tools such as handheld units.

Handheld units are fundamental for field services such as installation and maintenance procedures. They can be linked to other systems and devices, such as GPS systems and infrared readers. These interactions reduce installation time and reduce risks of unnecessary duplicated commissioning processes, which may be caused by devices installed in wrong locations. These handheld units are important not only for HAN devices but also for interacting with other infrastructure components in other network layers such as data concentrators.

Traditionally, these devices communicate using a local metering HAN, based on traditional ports such as flag ports or cabled networks. Media may not be a problem since they are adapted to the challenges of this new smart metering environment. Of course, these ports may have been built according to different needs, and associated specifications may not be compatible today or for future environments in the device life span. For example, these ports should not add additional risks of data security and tampering.

If the utility already owns assets on the network, it may be an appropriate option to adopt the media. An alternative option would be to use the local metering HAN for this purpose. It can generate several benefits such as reduction of local system complexities and costs related to additional communication ports.

In this scenario, the handheld unit would be simply another device in the HAN, but with temporary data access rights and different access authorization levels. In this case, the handheld unit should always be presynchronized with the middleware, then import the task, execute it in the field, and resynchronize it with the middleware for feedback. This synchronization can also be executed in the field. These units may be equipped for WAN communication, so for both daily operations and for exceptions, the administrator may easily generate new tasks for the technician to execute in the field.

Independent of the utility choice of strategy for media maintenance, middleware will always manage all the tasks of the smart metering environment in order to ensure synchronization settings in the field and other aspects such as data security and tamper proofing. Problems of synchronization between middleware and device settings in the field can cause high-impact risks for smart metering systems and must be avoided. The same applies for tampering and data security breaches.

***Smart meters.*** Smart meters are the next generation of meters. Even if part of the home intelligence is attributed to the in-home metering gateway, meters in the home will continue to be smart. They will continue to be efficient sources of measurement information in smart metering systems. Customer display units are optional in some architectures, and smart meters can also play a role as customer displays. In this case, the display and design must be adapted to customer needs.

New user-friendly designs are fundamental for smart meters to enable customer interactivity. This interactivity is key to enable new opportunities for utilities. Of course, the new designs must take into account all the specifications, tests and certifications regarding safety, metrology, and other criteria. For example, Yello Strom, a wholly owned subsidiary of EnBW Energie Baden-Württemberg AG, which itself is held in part by EDF Group, has been deploying an innovative and modern smart meter (in terms of design and technology) in Germany (fig. 2–32).

Fig. 2–32. Example of a new smart metering design in use in Germany. (Photo courtesy of Yello Strom.)

In terms of customer information feedback, monetary information seems to be more interesting or useful than energy units. Dual-fuel information is also essential for some markets. It may not be necessary to duplicate advanced displays in the home. For some applications, the electricity meter, being AC supplied, can act as a single customer display unit for dual-fuel applications. In this case, electricity meters would have a larger, multilined, multifunctional, and user-friendly display, while gas meters, heat meters, and water meters would have a basic and low-cost display design.

Meters must also be smart to self-manage some real-time functions related to safety, metrology, and other aspects. A gas meter, for example, should not allow gas use if there is a leak or if an unexpected open valve is detected, such as on a cooker. In this case, the gas meter itself must be intelligent enough to immediately interrupt the gas flow without any dependence on the HAN or decision made by the in-home metering gateway. The same applies to the electricity meter for reconnection procedures, but also for quality analysis of electricity supply. If an overvoltage is detected, for example, the electricity meter must manage it and interrupt the electricity supply if required.

Self-diagnostics, event management, and alarms are also important. Clearly, the meters must be intelligent enough to generate alarms and report them to smart metering systems according to predefined profiles, sometimes immediately when detected and others when queried by the in-home metering gateway, depending on the priority.

The smart meter should also be intelligent enough to operate in both credit and debit modes when required. A utility requirement for special debit meters generates additional costs such as meter acquisition and replacement. In a world where rates are more and more flexible, meters can give intelligent support at lower cost.

It is also crucial to analyze meter self-consumption at different levels. Beyond their normal burden, these meters can now power communication modems and switches that also consume energy. Different levels can be tested and specified, including consumption of modems on standby, in communication, when transferring data under interference, and when mesh hopping.

Whatever the customer segment—residential, industrial, commercial or any other—smart meters seem to be essential. Their intelligent functions are responsible for enabling many features and opportunities needed for these deployments.

**Customer displays.** This is the layer for the customer relationship (fig. 2–33). Customer displays are not necessarily physical devices owned by the energy utilities, but can also be incorporated into existing equipment owned by the customer. The ultimate goal is to allow customers to interact with services provided through smart metering systems.

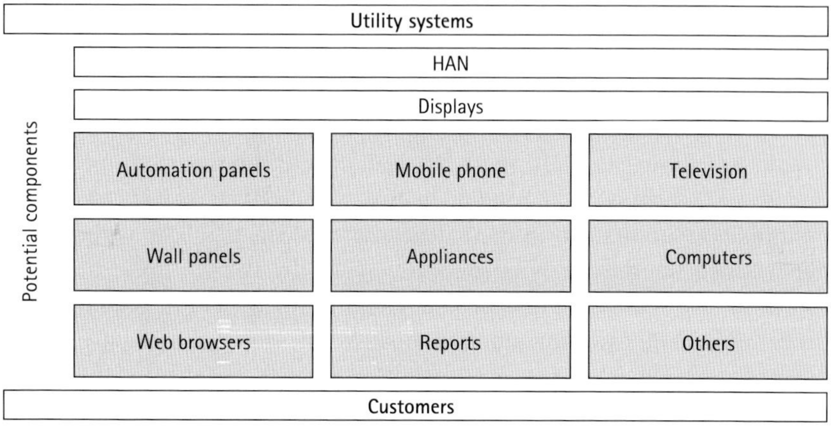


Fig. 2–33. Architectural environment diagram

Customer interaction is essential, so it is important that customers want to interact with smart meters. It would be even better if this interaction happened instinctively, voluntarily, and positively. Why not use existing equipment in the home or in the customer's daily life? Customers interact daily with some equipment such as televisions, mobile phones, and computers, which can also be used for smart metering applications. This would save money and would also be a guarantee of at least customer interoperability with the tool acting as a customer display. For example, EDF Energy, a wholly owned subsidiary of the EDF Group, has been deploying a television-based display on a smart metering trial. Customers can interact, using their TV remote controls, with this user-friendly application to better understand and manage their consumption.

Utilities can also use a customized display unit (fig. 2–34). There are several opportunities here, although this decision can entail costs, not only for technology acquisition but also for other aspects such as marketing, research and development, surveys, and education campaigns. Using an existing platform often generates costs related to integration, development of customized applications, marketing campaigns, and the like. Review of a full business case is recommended in order to cover the risks and benefits of this decision per utility and per market segment. The decision for using an existing or new platform varies from utility to utility and market segment to segment.

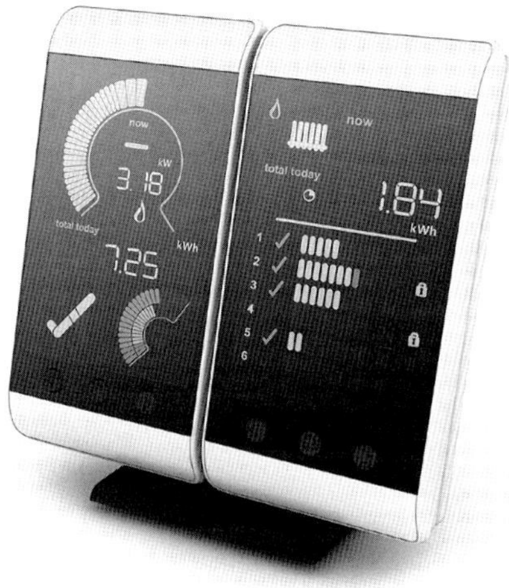


Fig. 2–34. Example of a customized display device. (Photo courtesy by Green Energy Options.)

When defining the customer display in a smart metering architecture, it is necessary to take into account the opportunities afforded to it. Excluding technical, financial, and other aspects, some customer-associated questions seem to be essential to define a customer display such as:

- How should information be displayed to the customers? Graphically? Synoptically? Textually?
- Should audible and textual alarms be provided?
- Should we have special interfaces for special customer needs, such as displays for sight-impaired people?
- Should it support financial transactions?
- Is online interaction necessary? If yes, via text chat or video conference?
- What are the customization needs? How can consumers configure its appearance and behavior to their specific needs?

- Will this be used as a marketing tool? If yes, by the utility only or also by partners?
- Should it be used to manage other devices?
- Should it be used to manage applications?

The management of requirements, opportunities, costs, and other elements, as in any project, must be prioritized, including what will not be in scope. Customer displays must support these requirements.

A display may provide feedback on achievement of energy efficiency and carbon emission goals. This is possible due to its ability to provide statistical information, alarms, and forecasts in different formats, such as monetary and graphical. It may assist customers to better understand their consumption, budgets, and goals for rational use of energy.

Displays can also provide in-home automation commands. This could be important for rate services such as heating control, load shifting, and home automation.

There is a trend toward dual-fuel rates, and customer displays are essential to help to understand both fuels. Interaction with both meter types can bring several opportunities. Opportunities for remote management of devices are becoming more interesting to customers. An example is the ability to securely and safely manage loads via web browsers. This could be provided on-demand or via scheduled services. This interesting element of climate control options for smart homes can be provided via a smart metering system. There is clearly a relationship between smart metering systems and smart home platforms.

There are also some options for in-home metering gateways offering web server management. In this case, customized pages can be provided to give customers local online information through Transmission Control Protocol /Internet Protocol (TCP/IP) using a normal browser and a HAN dongle device. It may be a good option to integrate this with private demand management platforms.

The number of new services to be offered seems almost unlimited because, with technology evolving, new displays appear every day, and each display seems to be more appropriate than the last one. A package of hybrid display solutions is recommended for a technical architecture. For example, Internet-based tools allow more customer interaction. Websites are sometimes better display tools for complex information such as graphics and analysis. Paper or electronic reports may also be good options for targeted customer feedback. Display devices can be a good option for online information such as rate changes, online energy monitoring, and in-home controls. Computer-based widget or messenger tools are also beginning to provide online interaction with customers. It is common for people to use these tools for online information about weather forecasting and public transportation or to communicate with friends. Smart meters could soon be another service for friendly interaction.

Finally, there is no best solution, just the most appropriate package of display options for the needs of a particular market. There are no limits to these services and opportunities.

## Communication Technologies

A good choice of communication media can help to achieve the goals of an energy utility, but a bad choice may produce serious consequences for the project scope, opportunities, and performance. Therefore, we need to discuss some of the concepts behind communication technologies.

Media deployment can be regulated by local telecom authorities as well as by regional and national laws or rules. The right technology use can vary according to several aspects such as country or regional regulations and concession agreements of energy utilities. Similar to standards, for example, the frequency of the signal to be used by a narrowband PLC in Europe may be different from that in the United States. It is important then to study all the applicable laws, standards, rules, regulations, and other legal aspects before deciding on implementation of any media. Some services, architectures, and technologies are subject to local laws and regulations, and companies must comply. Similarly, patents and intellectual property rights must be respected.

There is no perfect network technology, only the one most adaptable to your needs in a specific region. A hybrid communications environment is often advisable (fig. 2–35). Different products may be deployed for different customers. The scope of targeted services could be different depending on customer segment or deployed area.

In the architecture analysis, we mentioned some communication media for WAN, LAN, and HAN. Now it is time to discuss some of these media in more detail. There is a great deal of available detailed literature on telecoms, so the idea here is to discuss the subject at a high level only, in order to help you to understand the features of some telecom technologies. The objective of this section is not to cover all the existing media that could be used for smart metering applications, but only some of the most common.

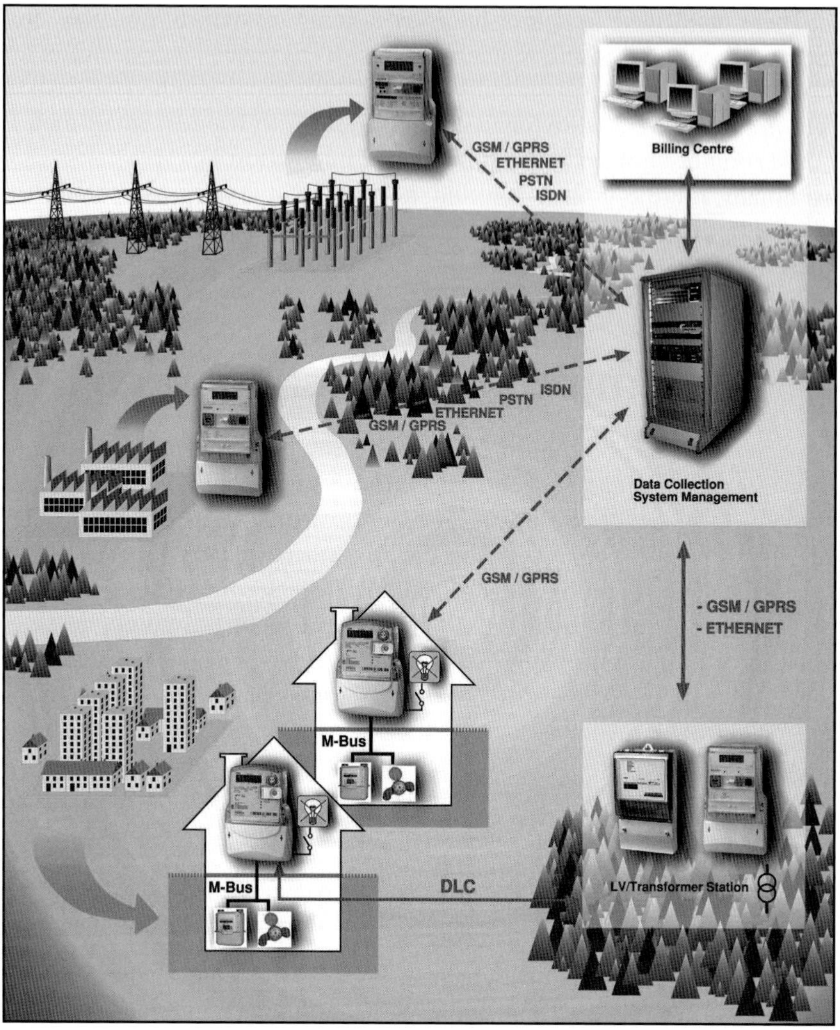


Fig. 2–35. Example of architecture based on hybrid media. (Illustration courtesy of Iskraemeco.)

## Power line carrier (PLC) technologies

PLC technology uses the electricity grid to transmit data. It superimposes a high-frequency signal on the existing electrical signal of 50 or 60 Hz. This superimposed signal is transmitted in kilohertz or megahertz. It can be propagated in an electrical cable and decoded by receivers. All receivers in the targeted electrical cable can then read the signal.

There are open and private international standards available for this purpose. As standards may directly influence the success of implementations, the choice of the most appropriate requires a serious technical analysis by energy utilities.

This technology can be used by utility applications but can also be used in the home, although the signal frequency may be necessarily different depending on the standard used and market rules. International standards are available for the use of technology in both environments. PLC technology can transmit data over low- and medium-voltage energy networks.

Electrical grids were built to deliver energy, not telecom signals. It is an unnatural environment for data transmission. There are several factors to take into account such as the number of spurs to supply homes, network impedances, grid reconfiguration events, and interference caused by electrical appliances. These electricity grid features may cause problems for data transmission such as attenuation, noise, and feedback. In the customer's premises the environment may also cause similar problems. As in any other telecom technology, measures need to be taken to minimize data transmission problems. Manufacturers provide research and development resources to optimize the performance of their technologies; performance can therefore vary from constructor to constructor. Because of this, as with all other telecom media, an efficient technical evaluation process is necessary. Laboratory and field tests are essential to evaluate these technologies and to ensure achievement of utility objectives.

**Narrowband PLC.** There are many versions of this technology on the market, using different modulations. Bandwidth varies according to the modulation technology used and implementation strategy. It varies from a few hundred bits per second (bps), such as versions using frequency shift keying (FSK) modulation, up to hundreds of kilobits per second (kbps) in new versions using orthogonal frequency division multiplex (OFDM) modulation. As bandwidth and speed of communication may not necessarily be dependent, these narrowband technologies are sometimes adequate for the needs of an energy utility in terms of the services they would like to offer their customers.

Performance can also be optimized by using different IT techniques such as data compression. Using this method, some narrowband PLC systems are able to significantly improve performance of data transmission. These technologies require data concentrators (fig. 2–36) that concentrate the tasks in their LAN. They provide the connection from the WAN environment to their managed devices. There may be WAN costs with data concentrators. Even if narrowband PLC uses the electricity cable infrastructure, data must still be transmitted from data concentrators to middleware (fig. 2–37). This may involve costs if, for example, mobile technology is used, although there are other options.

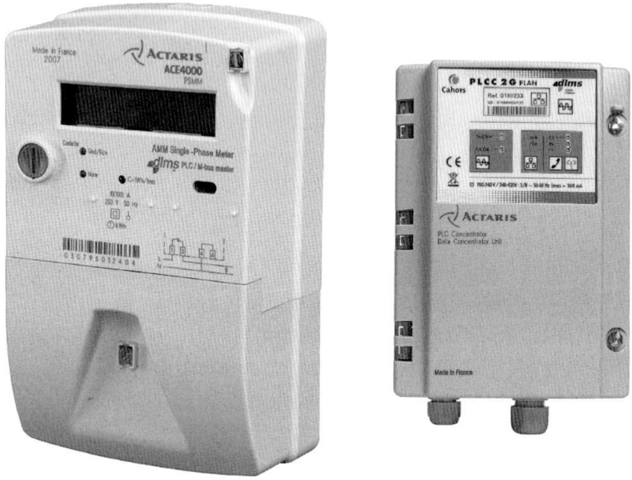


Fig. 2–36. A meter using a PLC narrowband and data concentrator. (Photo by Ludo, courtesy of Actaris.)

It is common in some utilities to have an existing communication link at their substations. In this case, this infrastructure may be used to reduce associated costs. Another option is the use of PLC technology over the medium-voltage network. In this case, special coupler devices are used to provide a physical connection to the network. They may be capacitive or inductive. In this case, as high-voltage substations are sometimes used as a corporate telecom infrastructure, the deployment may be executed without any permanent costs for data transmission in the WAN.

Data concentrators are essentially pooling mechanisms. Based on this, data concentrators generally always initiate the communication process, although some constructors have implemented mechanisms to enable push communication for diverse tasks such as alarm management and automatic detection of devices during installation processes.

Data concentrators are often installed in each distribution transformer, but in low-voltage interconnected grids, with a cluster of a few transformers, one data concentrator may be able to cover more than one transformer. The quantity of nodes covered varies depending on factors such as grid topology, distance between nodes, quantity of supply spurs per transformer, technology modulation, and implementation method. It may vary from several units to hundreds of nodes. Data concentrators are generally installed in substations, directly connected to distribution transformers, although due to the inconvenience for utilities during the installation process, some manufacturers have also developed solutions that enable installation of data concentrators at the meter level.

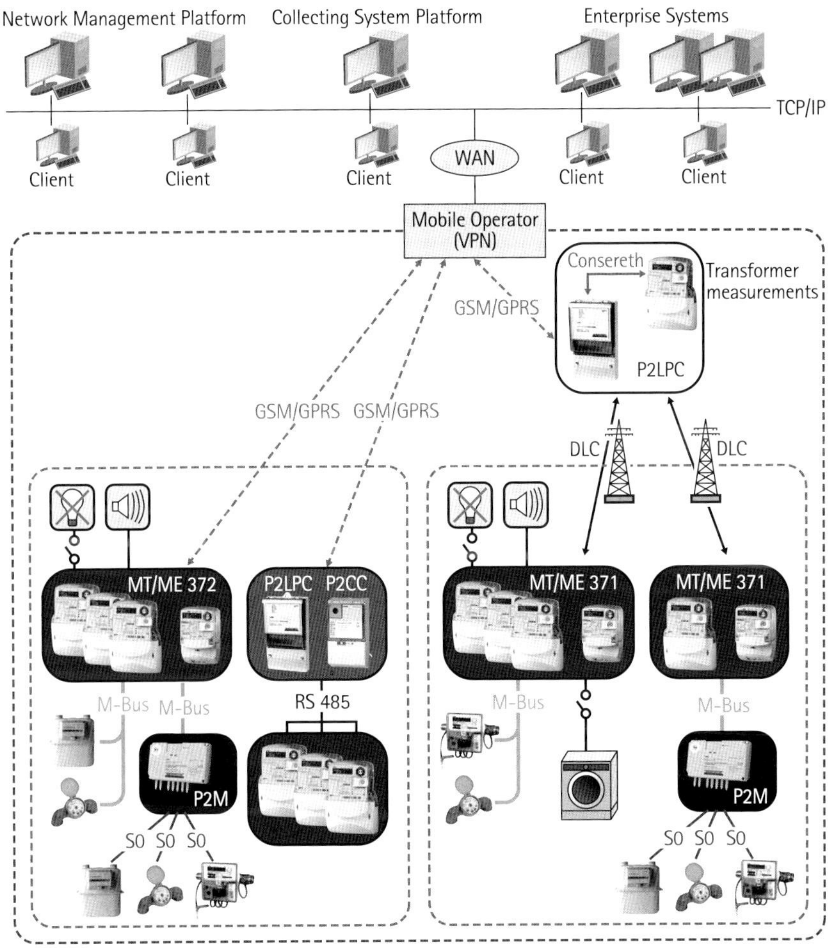


Fig. 2–37. Narrowband PLC architecture. (Illustration courtesy of Iskraemeco.)

Generally, the electricity meter (see fig. 2–36) itself acts as a repeater in the grid. However, sometimes, due to problems such as impedance and distance, it is common to use additional repeaters in the grid.

In terms of interoperability, gas meters are generally managed at the in-home metering gateway level, via the use of other HAN media. Another option is the use of bridge devices, as discussed previously.

Among the advantages of using this technology are the use of the existing electricity cable infrastructure of electricity utilities and signal propagation along the entire grid once the technology is deployed, excluding exceptions, of course. The first point is interesting in terms of cost of installation and maintenance. The grid is already an asset maintained and managed by the electricity company. In this case, the impact on telecom management may not be as big as when deploying other technologies such as low-power RF. In relation to signal propagation, the grid generates an opportunity to share the smart metering infrastructure with other projects such as smart grids. Energy network devices in the future may be deployed and managed by data concentrators. Other advantages are less dependence of the energy utility on telecom providers and signal penetration independent of building construction.

As with any other technology, narrowband PLC may introduce some disadvantages to energy utilities, such as the inability to offer differentiated services in competitive markets, the need to maintain telecom infrastructure, and problems of data traffic interruption due to interference. It is also extremely important to identify problems of interference with existing devices in the home. For example, test results have demonstrated that some types of in-home PLC technologies potentially interfere with the way light touch sensors work. A touch sensor is a device that turns on and off and changes lighting intensity with a touch on its surface. When both technologies share the same environment, this kind of light sensor may not behave well. Among other anomalies, PLC may change the intensity and status of the light successively without any human intervention.

**Broadband PLC.** This technology differs from narrowband mainly with respect to its bandwidth and signal. While narrowband technologies use hundreds of kilohertz to propagate their signals on the grid, broadband technologies generate signals in megahertz. Bandwidth may often vary from tens of megabytes to hundreds of megabytes. As with narrowband PLC, broadband PLC can be used inside and outside the home under specific standards (fig. 2–38).

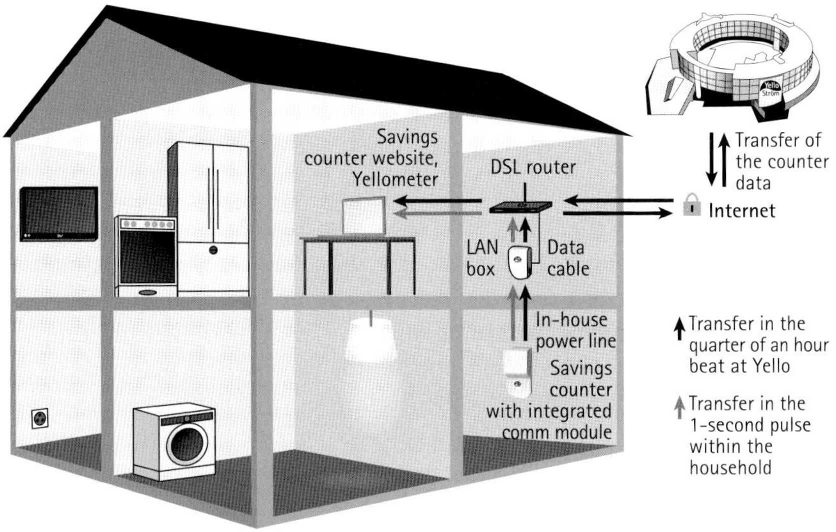


Fig. 2–38. Example of product using broadband PLC to communicate online with middleware through Internet modems of customers. To achieve this, a broadband PLC modem is embedded in the electricity meter, and another is connected to the Internet modem of the customer. (Illustration courtesy of Yello Strom.)

Bandwidth can provide an opportunity to electricity utilities to offer Internet services to their customers directly or via a partnership. Some energy utilities around the world offer related services such as voice over Internet Protocol (VoIP) and video offerings. This bandwidth also opens up opportunities for new customer relationship services such as video conferencing and data download via the in-home metering gateway.

Since the start of broadband Internet deployments around the world, the "last mile" has always been a problem for telecom providers. There are many issues involving the physical deployment of this medium. Rural deployment of broadband Internet is one challenge for telecom companies that broadband PLC may be able to answer; however, sufficient cable infrastructure may not be installed in customer premises. Another challenge is that wireless modems do not always provide coverage to all parts of the home. This is inconvenient for clients and may affect the market share of telecom providers.

Based on these problems, telecom companies around the world have studied the benefits of broadband PLC to provide their last mile. This may result in beneficial partnerships between energy utilities and telecom companies with common interests. Utilities benefit if they do not have the telecom expertise and would like a minimum WAN bandwidth availability to be used by their smart metering and smart grid deployments. Telecom providers would like to profit from the cable infrastructure to deploy efficient in-home, last mile, and rural media. Partnership arrangements between both entities may depend on legal aspects. These arrangements must be carefully studied case by case.

The infrastructure generally uses head-ends to provide connection between devices and the WAN. This uses capacitive or inductive couplers to provide physical connections to the network as well as repeaters to improve signal penetration and performance. Depending on laws, standards, and strategy of deployment, this technology may also offer an opportunity to use broadband PLC inside and outside the home. This may reduce the number of additional modems, depending on the deployment strategy and services the utility would like to provide. It also offers an opportunity for interoperability between smart meters and computers due to the use of TCP/IP in the majority of implementations.

This technology, as in any telecom technology, offers opportunities but also risks. Some opportunities have already been discussed. Some of the risks are in the grid environment and the fact that the infrastructure needed to ensure the required level of broadband services to customers may not be cost-effective.

## Low-power RF technology

A mature technology that is used around the world, low-power RF devices are generally organized in a mesh or star network to provide infrastructure to smart metering systems.

There are several manufacturers that offer RF chipsets using different frequency bands of radio operation. These bands are generally not licensed for the majority of countries, although there may be telecom regulations and laws on frequency band allocation, transmission power, or other criteria of a specific region where an energy utility is located.

These RF nodes normally require the use of data concentrators. As in narrowband PLC, they are responsible for concentrating and managing the data-related tasks in the LAN. The number of data concentrators per location may vary according to different factors such as technology, chipset implementation, signal propagation range, interference in an area, and building topology. Because of this, an efficient evaluation process is generally necessary in each area of deployment and type of technology. As material penetration varies by frequency band, specific technology may be better applied in one location than another.

Depending on the distance between customers, sometimes the cost of data concentrator devices can be shared among nodes. For example, assume that a data concentrator costs 200 dollars and nodes cost 20 dollars, so in an arrangement of 100 nodes, assuming the need for only data concentrator and the meters, the cost per node would be 22 dollars. In another scenario involving only 10 meters in the coverage area of the data concentrator, this cost would reach 40 dollars. Because of this, some solution providers developed low-concentration data concentrators. They provide the same range of radio coverage but less processing capability and functionality. Their cost is usually lower than a conventional data concentrator and may be an interesting solution in rural arrangements or areas where the distance between customers is outside of acceptable levels of nodes concentration.

Another procedure generally used to improve the concentration level of these solutions is to optimize the position of data concentrators and antennas. Sometimes, however, this is not possible due to, for example, use of nonutility premises, urban building topology, the absence of places to strategically install it in a targeted premise, and extra costs for rental of places to install these devices. Given the difference between theoretical plan and actual practice, these costs should be included in the business model.

We can reduce dependence on data concentrators with mesh topology, which provides additional hops in the RF network. These hops may use several nodes to communicate, depending on the technology used. Mesh topology expands the coverage of the network, although the addition of extra hops may imply a reduction in RF network performance, depending on how they are organized.

These networks provide several opportunities. As with narrowband PLC, there is the possibility of future integration with other projects such as smart grids. This may be an interesting future-proofing feature for utilities. Another opportunity may be to use mesh RF media for both last mile and HAN environments. This can eliminate extra complexity and costs of extra local media in the in-home metering gateway and its HAN. In addition, problems related to interference and the building environment may affect performance. As in all implementations, technical and financial studies must be done case by case.

## Mobile

A variety of mobile technologies are available around the world, such as time division multiple access (TDMA), code division multiple access (CDMA), and several others, although the most common for smart metering implementations is currently GSM. For this reason, this discussion focuses on GSM. Of course, any other mobile technology could be used in your deployment using similar principles. The choice of the most appropriate one depends on evaluation processes in the energy utility.

This solution provides WAN directly to in-home metering gateways (fig. 2–39). Telecom providers often own the entire telecom infrastructure. These networks offer opportunities for mass rollouts, but the medium also presents an interesting opportunity for segmented rollouts because it does not depend on new concentrated infrastructures. This may be why energy utility pilot projects around the world try this technology first to understand the behavior of their customers in different areas and segments.

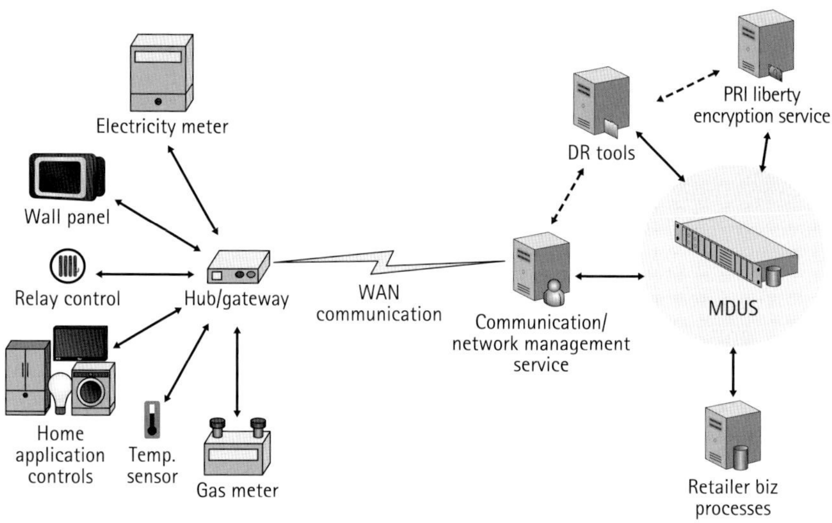


Fig. 2–39. A product using a GPRS metering gateway. (Illustration courtesy of PRI.)

Mobile operators usually offer VPN (virtual private network) arrangements for energy utilities. In this case, a secure link is established between the utility's middleware and its gateway devices in the field.

Because the telecom infrastructure already exists, devices can be installed more quickly in the field. Coverage may still be an issue in some countries due to the fixed location of devices in a customer's premises. If your mobile phone does not work in a specific location in your home, you can often move to another room for better reception. This is not possible for meters; once installed, they usually stay in the same location. Contractual arrangements ensuring coverage and the use of external antennas are some of the procedures used to mitigate reception problems.

Another point is the dependence on a single mobile operator. Multioperator subscriber identity module (SIM) cards may help to mitigate this problem. Roaming services can also be another interesting solution. These arrangements may depend on telecom market rules and need to be studied case by case.

Communications technology life span is another disadvantage. As with any technology, it can become outdated. This kind of issue can also be reduced by well designed contracts with telecom providers and good evaluation processes. This risk may also provide an opportunity to energy utilities. When a new mobile technology is deployed, telecom providers may offer contractual arrangements to enable the use of old technology for a defined period at cheaper costs.

Often when these technologies become mature, the infrastructure has already been partially or fully paid for. Because of this, lower prices are being offered by telecom providers to encourage continued use of these technologies. Another cost reduction factor is concentration of the majority of tasks in off-peak periods of data traffic. Agreements setting fees related to data volume instead of per node may also contribute to savings. Taking into account the level of savings offered and that there is no requirement to maintain these networks, this could provide a cost-effective solution to energy utilities for mass rollouts.

This technology can also be used for WAN connectivity to data concentrators. It is important then to evaluate this technology as an isolated medium as well as a WAN medium for other concentration-based technologies.

As observed, there are several points to be considered in a business case analysis. Because technologies are not easily compared, we need to define efficient technical and financial comparison criteria when performing business studies or evaluations.

## Internet broadband piggybacking technologies

As with mobile technologies, this technology can be used not only to connect isolated points directly via WAN but also to provide WAN connectivity to data concentrators. Although some technologies such as mobile and broadband are able to offer broadband Internet, the objective of this section is to consider other technologies that may offer the same capability, such as cable Internet, xDSL (digital subscriber lines), satellite, and others. There may be additional opportunities to consider with these technologies.

An interesting point, as briefly discussed in other sections of this book, is the possibility of using the customer's Internet to provide WAN connectivity to in-home metering gateways. Although we have previously discussed disadvantages with the use of this technology and restriction of services it may cause, it is still an interesting option to deploy in some cases. There is a trend for other solution providers in nonenergy markets to use it.

A key point of Internet piggybacking technologies used by customers is customer choice. They should not just agree on the use of the Internet but also realize there is added value from the services to be offered to them through this medium. In addition to legal agreements, this seems to be the most efficient way to ensure that this medium is always available for energy utilities to use.

It is also important to set boundary responsibilities around the use of shared media, such as data security and faults. For example, a customer may blame energy utilities for faults caused by their equipment or breaches of data security. Responsibilities have to be clearly defined.

Sometimes, ironically, energy-efficiency advice may become a problem for utilities. For example, following utility energy-efficiency advice, in order to avoid standby load, some customers commonly disconnect their in-home Internet modem at night and when away from home. This procedure may produce telecom service interruptions to smart metering systems. Action is required to mitigate this risk, such as efficient data recovery processes, education campaigns, and legal agreements to be able to offer some online services based on clear rules and predefined limitations.

As people become dependent on the Internet, more Internet services are being deployed around the world. In some countries, the majority of households are already covered by broadband services. Clearly, there is a trend that most premises around the world will soon be covered by some source of broadband Internet service. This trend and the absence of piggyback communications for customer services make this medium very attractive despite its disadvantages.

Another opportunity is to share the use of wireless infrastructure provided by these modems, such as Wi-Fi. This may yield a low-cost solution to utilities, as modern modems usually offer this capability. Of course, there is a setting dependence, because the gateway must know the security access key (e.g., wired equivalent privacy, or WEP) used by the customer. Another alternative to nonwireless modems, or to avoid this setting dependence altogether, is to opt for dongle bridge devices. These are generally connected to the existing Ethernet port of Internet modems. By using the Ethernet port, connectivity is provided to the in-home metering gateway using technologies such as in-home broadband PLC. Despite the disadvantages caused by physical constraints, another alternative is to use Ethernet networks to connect the in-home metering gateway to the Internet modem.

## Public switched telephone network (PSTN)

PSTN is possibly the most mature method of communication for metering systems. Utilities first began using this medium to provide telemetry services to enable communication to their private meters in energy networks and for industrial customers.

Implementation of PSTN requires modems or similar devices connected to or embedded in the electricity meter. Modems are installed in field devices, and a group of modems is installed by the utility. Data are transmitted from the field to the utility via modem-to-modem connections. Sometimes, the existing line of the customer is used. The utility then establishes one or several time windows, for example, twice a day, to collect meter data. The advantage is that the PSTN is using the customer telephone line for "free."

Disadvantages could arise, such as the number of modems to be installed in a patch to enable frequent data collection, frequent communications interruption, and the absence of online communications. In an event of an alarm, for example, communication problems may be an issue.

Another form of connection is the use of Internet dial-up processes for data transfer. The disadvantage is that this connection is susceptible to all the known constraints of dial-up connectivity.

Because of these issues, a common view is to regard this PSTN as obsolete, but this is not true. For example, remember that xDSL technology uses this medium to provide Internet broadband to customers. Several other technologies may be used for the same purpose.

Despite the strong penetration of mobile networks around the world, PSTN still provides service to energy utilities and generates interest by telecom regulation authorities in several countries around the world. Some authorities are conducting studies into future international in-home interoperability. One of these studies looks at the use of these media by nontelecom companies in a fully competitive environment. This could be an opportunity for smart metering deployments and several other in-home services.

Another common use of PSTN is coverage exceptions. Thanks to strong penetration, PSTN lines are almost everywhere and can efficiently cover remote exceptions. This could apply to data concentrators and isolated nodes.

## Licensed RF

RF broadcast technology is often able to broadcast a signal for several miles. There are various radio technologies available around the world with different telecom features. One example is use of very high frequency (VHF) technology. The first versions were implemented as one-way technologies and offered only a few bps of bandwidth. New versions offer two-way capability and up to tens of kbps of bandwidth.

A clear advantage of broadcast technologies in relation to others is coverage. A few telecom masts may cover an entire energy utility area. Another is that this medium may act as a WAN, LAN, and HAN. Depending on the targeted services, it could enable in-home services without additional HAN devices. It also provides an opportunity for sharing telecom infrastructure with other networks such as smart grids or smart metering for water companies.

Potentially covering the HAN, RF broadcast technology is often unlikely to be compatible with devices such as gas meters and in-home appliances. This incompatibility is due to several factors, such as trends in home automation, the high transmission power of this technology, and high consumption of these devices. Another aspect is the size of antennas for these devices and sometimes the devices themselves. They may be incompatible with existing applications and built-in designs and require the use of separate gateways.

This produces a centralization of processing of all devices in middleware management. This may be compatible with some services but may be an issue for online and real-time local services.

Despite the benefits in a private network, such as legal freedom from interference due to temporary band concession and opportunities afforded by sharing the infrastructure for competitive energy market environments among different energy suppliers, there are disadvantages with the management of telecom infrastructure. Because this management is more complex than the management of low-power RF systems and requires more specialized people to manage it, these kinds of deployments are normally executed in turnkey arrangements that generally include network maintenance.

## Emergent RF telecom media

Telecom providers are actively studying RF telecom media to upgrade existing mobile and broadband Internet media. Several telecom authorities are providing more consultations, test authorizations, and concessions for telecom providers to use these technologies.

These technologies may offer an opportunity for in-home metering gateways to be used as in-home Internet hotspots or femtocells, which are small cellular base stations used by telecom providers to extend the use of Internet indoors. These potential applications may enable partnerships between telecom companies and utilities that have common points of interest.

Two examples of emerging media are worldwide interoperability for microwave access (WiMax) and LTE (long-term evolution). These media could potentially be used in smart metering deployments.

## In-home RF media

Some of the media options discussed here may also be used for in-home applications (e.g., HAN), although there are others that are often used and designed for this purpose. Examples of these are Z-Wave, ZigBee, Bluetooth, Wi-Fi, and several other private and open technologies.

Our objective here is not to compare these technologies. This is a task for utilities in their evaluation processes. Instead, we describe some of the issues and advice for deployment of RF technologies in an in-home environment.

The home can be a difficult environment for RF media. Because of this, it is important to ensure that the selected medium is able to communicate in the majority of an energy utility's installation environments.

It is important that signals be able to penetrate not only the materials inside the home but also those used in its construction. To accomplish this, it may be necessary to propagate the signal from inside to outside the home and vice versa, depending on the location of the meter. Thus, an electricity meter could be located inside the home and the gas meter outside. Different RF technologies can provide different penetration levels, depending on the materials they face in the home. Examples of these materials are metal objects, structures and frames, concrete walls, double-glazed windows, and roofing shingles.

Sometimes meters are installed in metal-framed cabinets, and special solutions need to be considered in order to avoid the signal-blocking effect of a Faraday cage. Another point to consider is interference. Several pieces of equipment in the home can cause interference to some RF technologies, such as microwave ovens and broadband Internet wireless modems. Coexistence is also a fundamental consideration, not just to avoid being affected by other networks but also to avoid interference with other devices in the home.

There are several solutions to mitigate these problems, such as the use of low-noise amplifiers, mesh capability, and automatic changes of frequency channel. The chosen solution must be implemented according to regulatory rules and standards. Legal aspects may also apply to some media in some regions, as well as regulatory rules if there is communication from one home to other.

Examples of criteria to take into account for in-home media evaluation processes are:

- Plug-and-play installation capability
- Presence of international standards
- Methods to deal with interference, such as multichannel capability and related automatic channel change
- Effective speed of transmission
- Transmission range
- Data security mechanisms
- Future-proofing
- Robustness
- Cost
- Maturity
- Use in smart metering implementations
- Use in other potential industries such as home automation or appliances
- Support for remote firmware upgrades
- Media-related energy consumption (especially important for non-AC supplied devices such as gas meters)
- Interoperability capability and future-proofing
- Data management methods

The majority of these criteria apply in the evaluation processes for any communication media. Many other telecom and nontelecom aspects must also be taken into account when looking for the most appropriate in-home communications technology for your environment, such as data transmission, safety, laboratory tests, legal, IT, and several others. A utility company must consider all these factors in order to perform an accurate evaluation.

There are opportunities for individual media. For example, Bluetooth is currently used for the majority of mobile phones. These devices can act as local displays with no additional communications costs, and they are used daily by consumers.

Several other challenges need to be addressed for in-home and other utility environments and must be reviewed in the selection process. This book is not meant to be a comprehensive reference on the subject of communication technologies, as there is much literature already available. Instead, it presents some of the challenges of this subject and stresses the importance of the media evaluation process as essential to the success of your smart metering project.

## Data Security

In traditional metering systems, meters were completely decentralized devices for measurement management. Together with billing systems, they were the most important element for billing customers. In smart metering systems, they are simply elements of a more global system. As with any element of the system, they are important and carry out their tasks, but the full achievements come from the system as a whole.

Focusing on the system leads us to look at enhancing the system globally, not just some isolated elements. For example, in the scenario with isolated meters, if someone wanted to interfere with billing, the easiest way was to try to tamper with the meter installed in a specific premise. If someone with bad intentions wanted to cause a local blackout, the best point of action would be a substation. In this new scenario, it is different. Device management is fully centralized, and it enables new risks of illicit system interference. Changing data could interfere with several meter registers at the same time. Broadcast commands generated by a hacker could also cause a blackout. This is a new world where metering is now part of an interconnected, computer-based environment, and this requires more attention and compatible action. This does not mean automation processes should not happen, but that they must be properly implemented. The financial world is an example. Billions of dollars circulate around the world in short periods even if treaties exist. Data security and validation policing must be implemented to ensure that transactions are done with the lowest possible risk. Similarly for metering systems, taking into account the different risks, proper actions must be taken.

Associated with data security, data validation is also key. Tasks should be controlled by intelligent algorithms, and incoherent tasks must be promptly treated, blocked if necessary, and properly investigated.

We do not intend to teach data security and validation here, as there is much other detailed literature available, but to draw your attention to this subject and the required attention it needs in this new metering environment.

It is necessary that metering specialists and IT specialists, who sometimes worked in completely isolation, now work together to achieve the common objectives of an energy utility's smart metering projects. It is also important that the company is fully engaged as a whole with this process because data security is not just related to data management tasks. For example, confidentiality is an important element. Managing confidentiality is impossible without a total company strategy and cooperation. This problem not only affects the utility, but it also may affect several participants in the energy market, because these are public services. Other participants such as local legal authorities should also be involved actively to help to mitigate these risks.

## Data security analysis

One of the first actions in any data security analysis should be to define the scope of the project. For example, what types of transactions is the system able to deal with, and what types will be excluded? Are they financial, confidential, or safety related? A well-defined scope is always a good first step in any project.

We should identify the threats in the environment we are dealing with. By taking the metering environment in isolation, rather than considering all the in-home and other integrations, it is easy to identify several threats. Among others, we could identify these:

- Hacker attacks in the several layers of the system such as IT systems, WAN, LAN, HAN, and in-home metering gateway and devices
- Cloned devices—a real risk for smart meters if we take the telecom environment as an analogy

- Confidentiality breach
- Unauthorized data access
- Reverse engineering
- Illicit applications and firmware interventions
- Mistakes made by untrained people or illicit tasks executed by ill-intentioned users
- Stolen devices

Several other threats may be also identified (fig. 2–40). These are just a few examples. Utilities must carry out efficient analysis with smart metering system providers and other participants in the energy market.

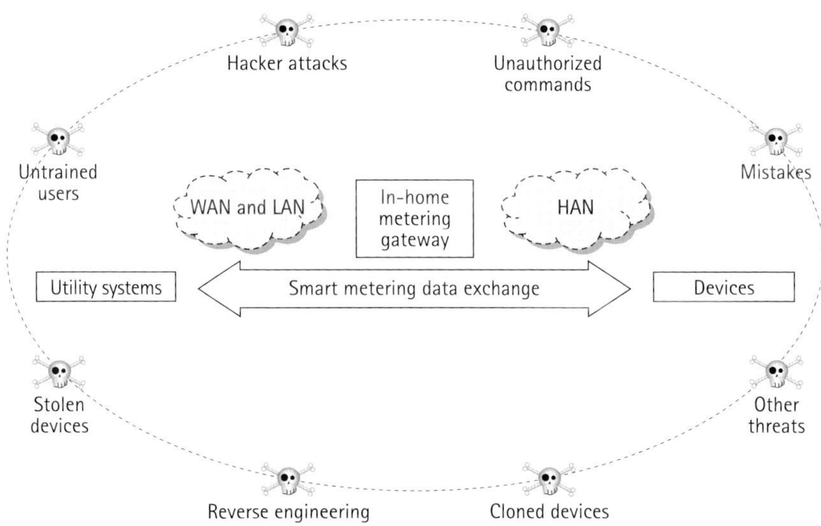


Fig. 2–40. High-level threats diagram

Once threats are identified, their associated risk must be properly mapped. Risk analysis and logs must be created and maintained with frequent reviews and updates. Threats and risks are not static, so reviews and updates are essential to an efficient process of evaluation. These risks must be analyzed, taking into account different levels such as safety, finance, and company reputation.

Corporate data security and confidentiality policies should be reviewed in order to comply with this new environment. Data security actions must be taken, such as VPN implementations, dynamic data encryption, intelligent algorithms for action to be taken in real time during attack events such as port lockouts, frequency hopping, efficient authentication, logs and alarms, virus detections and corrections, reverse-engineering shields, resilience to noise and interference sources, illicit device removal detection, illicit parameters, firmware and application update attempts, data integrity management, real-time monitoring modes, efficient maintenance protocols and ports, backup and data recovery management, clock synchronization, access management, and refinement of firewall rules. In summary, all the traditional IT procedures apply (fig. 2–41).

Related training and education campaigns are equally essential for the success of projects. This should be applicable not only for users but also for all who directly or indirectly interact with these systems.

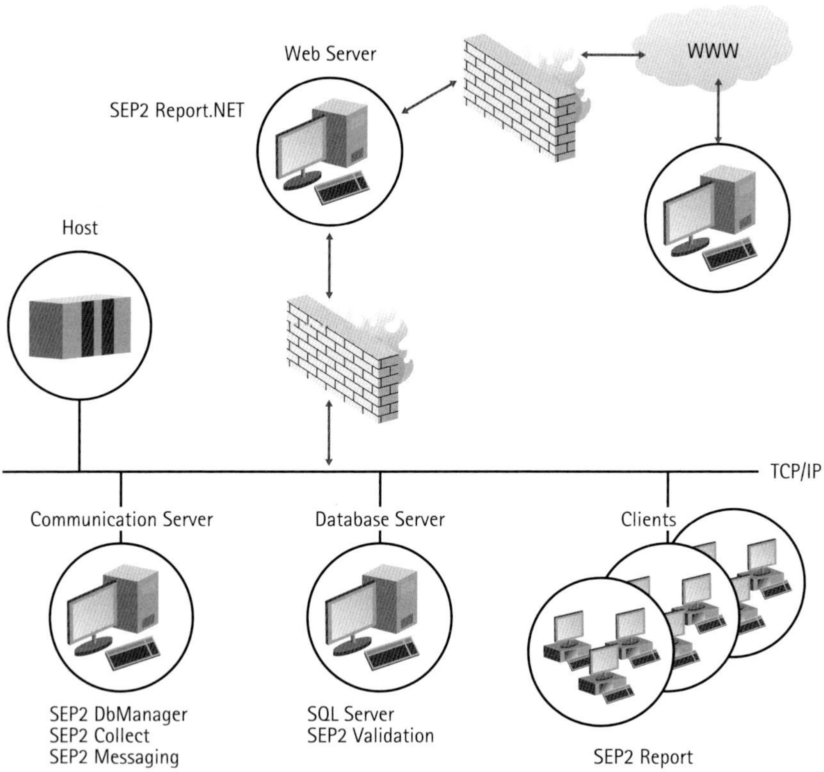


Fig. 2–41. A corporate IT environment. (Courtesy of Iskraemeco.)

Products should also be properly evaluated. For example, computing power and media technology should be appropriate to the required levels of data security. Encryption procedures must also be investigated, because solution providers may claim to implement encryption but in reality, they just encode data (data confidentiality is not ensured).

Another point is traceability, which is essential for identifying and mitigating detected problems. It is hard to audit and investigate what is not traceable. Transactions should execute only once, should be device-, task-, and utility-specific, and they should have a period of validity. In international metering standards, different parameters are used, together with proper security keys as part of dynamic encryption procedures. It is important to ensure that even if an encryption process is breached for one transaction, it will not be reproduced in other tasks.

The company crisis management procedure should be updated, taking into account this new environment. Utilities should address potential exceptions with prompt, predefined, and efficient action.

Action must not only be taken in the corporate environment and in middleware, but it also should be taken in all the different layers between middleware and devices. For example, the in-home metering gateways play an important role in data security and safety. Lockout actions may result from real-time tasks in these devices related to, for example, attack detection. Meters and other devices also have roles to play. For example, if abnormal behavior of consecutive switching orders is detected, the order must not execute and the anomaly should be immediately reported. If caused by external orders, a port lockout could be activated during a predefined time window, without waiting for any external commands. The end-to-end environment must be reliable, secure, and safe.

These are just a few examples of data security. Utilities, their partners, and other participants will also be involved to ensure that a secure and reliable environment for smart metering systems exists.

Finally, data security must be implemented in an end-to-end VPN basis, not just to the point of the metering gateway. It must be simple, efficient, well understood, flexible, and proven, but also innovative. It is thus a clear challenge to be properly investigated and efficiently achieved.

## Data and task validation

Taking into account the number of transactions smart metering systems generate in short time intervals, data integrity must be ensured. Intelligent validation algorithms are used to detect and correct irregularities and anomalies. Even with traditional data security procedures in place, several exceptions may still occur, such as errors and illicit behavior by users. Proper filters should be in place to help mitigate these potential problems.

Using the same example as above, consecutive orders of connection and reconnection of electricity supply in a short time period are tasks worthy of analysis and investigation in the middleware side, too. If the number of orders and elapsed time between them does not fall within defined parameters, they should not even be sent to the devices. Registers could also be analyzed, and inconsistencies should be identified and treated. These consecutive orders may be the result, for example, of a hacker attack or a meter fault.

An in-home metering gateway could be a source of intelligent validation. It should diagnose abnormal events and data and take real-time action. The devices themselves should also provide some validation. Validation is an end-to-end process.

Inconsistent data should not be managed together with real data in the system. Management tools must be dedicated to this analysis. For example, diagnostics must be managed by workflow tools to ensure that corrective actions are in place. For illustrative purposes, imagine a customer was incorrectly billed due to a fault occurring exactly at the end of billing cycle, and the bill was generated uncorrected. Several actions must take place. The old bill must be cancelled, a new bill promptly generated, and algorithm corrections must be made so that this problem will not be repeated.

With the advance of artificial intelligence, several universities and research centers are currently working on self-learning data validation algorithms. Because these validations usually vary from environment to environment, research and development partnerships with these groups are essential to developing efficient tools. Although these tools can help avoid the eventual consequences and risks of bad validation procedures, even with self-learning algorithms, nothing can substitute for human ability. Human resources must be engaged in this process. IT validation systems should only be auxiliary tools for people, not the opposite. Proper staffing of a smart metering project is paramount to its success. The people involved must be trained, motivated, and fully engaged with the project and company objectives.

## Data Processing

More important than the entire smart metering infrastructure are the benefits utilities can obtain from the best data processing. Because of this, it is common for these projects to accumulate vast amounts of data. This sometimes requires several IT resources such as applications, infrastructure, and people to manage data storage. However, large amounts of accumulated data do not necessarily produce benefits to the utility.

Isolated, data have no significance! Data have to be processed and transformed into useful information for energy utilities. Applying this information within the knowledge base allows the organization and professionals to become wise. There are several concepts and challenges behind the gradual process of transforming data into wisdom. The intention here is not to explain all these concepts, as there is much available literature on it, but simply to help you realize the importance of data processing and also that stored data must confer benefits to the company, not just occupy space and generate costs.

Thus, before discussing data processing, the utility should initially work on requirements and specifications related to data generation and data acquisition strategies. Being obsessed with data is sometimes a challenge to overcome in strategy-building processes.

Simple questions may help you to deal with these strategy definitions such as: What? Why? Who? Where? When? How? How long? How much? Based on this, there are several related questions such as:

- What kind of data are we talking about?
- Why do the data need to be generated by the devices?
- Who owns the data?
- Who needs the data?
- Is all the added-value data being generated by the devices?
- What financial and nonfinancial values may be associated with the data?
- Must all the generated data be stored? If yes, where? Locally at the devices? In a proper database?
- For how long do the data need to be stored in primary media?
- Does these data need to be archived and for how long?
- When will the data be needed?
- Are there any legal requirements associated with the data?

- What kind of information can be obtained from the data now and in the short, medium, and long term?
- What are the criteria of success related to data processing?
- Does the utility have enough internal knowledge and resources to process the data properly?
- What are the risks of processing or not processing these data?
- What is the cost of data processing?

Once the data generation and acquisition strategies are defined, it is necessary for the utility to process the data. A common mistake in data processing is thinking that nonspecialized people can do this work or that manual tools and instinctive analysis can generate expected results. These attitudes must be overcome. There are several types of professionals who can contribute to this analysis, such as statisticians, information managers, and information architects. They must participate in the smart metering team in order to help define tools and methodologies that will help energy utilities achieve their goals.

A clear example of value-added data is load profile. If correctly processed, the load profile should help utilities to understand their customers' behavior. This may help customers to achieve better results in several areas such as rational use of energy, coordination of their budget with their consumption, efficient use of microgeneration methods, and rate advice. Based on this, a visionary utility is also able to create competitive differentiation like new customized products and rates, efficient use of the different generation sources, achievement of regulatory goals such as supply quality and carbon emission reduction, reduction of nontechnical losses, process optimization, and reduction of operational costs.

In addition to individual analysis, there are also several methods of comparison related to different measurement curves. Based on the use of intelligent algorithms, customers with similar behavior may be easily grouped together. Individual profiles may also be compared to group trends and to standard profiles for different purposes. Logical online processing may also be used. The opportunities around this subject are almost unlimited.

In order to better understand these opportunities, let us define an environment to study. Assume an electricity utility installs a meter connected to a distribution transformer. Its data are properly acquired, along with every other customer's meter, every 10 minutes. Among several other purposes, online data analysis is generated for nontechnical loss evaluation. Suppose that every 10 minutes the system provides an online analysis between the energy supplied by this transformer and the sum of all related meters. Now we can build some scenarios.

In a first scenario, we assume that during the current month, the energy discrepancy obtained by the online analysis is on average 2%. Assuming this value is optimal for this circuit, if in the next interval the collected data show the discrepancy is now 5%, the system could easily generate several automatic investigations. Let us assume one would be the automatic comparison of all individual load profiles to a set of rules such as abnormal seasonal consumption reduction and comparison to customized standard curves. Assume that these filters show five customers with a high probability of being the cause of this anomaly. If one of these customers produced a meter fault or tamper alarm, this would immediately generate a field inspection work flow order for this customer and online monitoring up to the correction of the anomaly in the field. Assuming a parallel scenario where the meter reported a fault alarm and generated a field investigation. This scenario could also be analyzed at other levels like energy balancing on individual medium voltage circuits, substations, and total energy balance.

Assume that it had generated an inspection and the discrepancy disappeared during the field investigation. If the tamper alarm suddenly reappeared hours after the technician left the premises, it would deserve additional investigation for the treatment of a more complex situation. We may be dealing with an illegal measurement system manipulation.

Another interesting feature of data processing is the ability to cross-reference sources of data. For example, online load profiles compared with online weather measurements and forecasts could result in an efficient tool of energy management. This would be important for demand forecasts and avoiding energy blackouts, with preventive actions such as load curtailment commands, network reconfiguration, rate-related load shifting instructions, and distributed energy storage management. With mass deployment of smart metering and smart grid projects, it would be easy to imagine energy predictions and actions when online information was analyzed, such as energy requests in a spot market.

Data processing is also a continuous learning process. Analysis feedback is important to improve the accuracy of the rules and related data management tools. An intelligence center should be dedicated to this in different areas of the business such as marketing, field services, energy generation, business, and retail.

Technology advance is a reality and should be recognized by energy utilities. Utilities must invest in efficient IT systems and infrastructure for data processing purposes. IT teams must also actively participate in this process. Data processing is an art, and as with any art, it has its associated price. Efficient financial analysis is needed to measure the impact of investments and the potential benefits they can generate.

This is only one part of a vast theme; connected aspects not covered here, such as business intelligence management and data quality management, should also be considered.

## Interoperability Challenges

Interoperability is one of the biggest challenges to be faced by smart metering projects. It can help projects to succeed, but it can also cause them to crash. We have briefly discussed this subject in previous sections of this book, but what is interoperability?

Definitions of interoperability depend on the source. These sources include international experts who participate in projects and initiatives about interoperability, supported by different market participants such as governments, research entities, universities, solution providers, and utilities. Some specifically deal with smart metering and in-home challenges, such as The Application Home Initiative (TAHI) in the United Kingdom. Its mission is accelerating the adoption of connected homes. It aims to facilitate change and create alliances to ensure that smart homes are fully connected and will deliver beneficial services. TAHI defines interoperability as "the ability of two or more entities to communicate and cooperate despite differences in the implementation language, the execution environment, and/or the model abstraction."

TAHI's definition of interoperability demonstrates how big this challenge may be. To better understand it, the reader should understand what may *not* be defined as interoperability:

- Coexistence: the ability of two parties to exist together. In a smart metering environment, this could mean the ability of two or more devices, applications, or communication media to share the same environment. However, these differing entities may be unable to communicate to one another. Further, the operation of each device should in no way disturb the others.

- Interconnectivity: the ability of two different entities to connect one with the other. In this case, these different entities are more than to be able to coexist; they are able to communicate. However, this does not yet provide the required level of interoperability.

> It does not provide other features needed to achieve this level, such as the ability of cooperation between these different entities and all its related opportunities. Synergies may not necessarily be created by generating extra opportunities. Interconnectivity is then just one challenge to be faced to achieve interoperability.

Using these concepts, we can say, for example, that ZigBee and Wi-Fi devices are able to coexist on the same frequency of 2.4 GHz, but these devices are not able to directly communicate with each other. Similarly, two different Wi-Fi devices may be interconnected through a Wi-Fi router providing Internet access to their different applications. However, these different applications are not necessarily interoperable.

As with data processing, interoperability requires full investigation and analysis. An efficient interoperability strategy must be defined by energy utilities in order to set the scope of the project. It would not be logical to enable interoperability to every possible entity without analyzing the benefits to be obtained. Interoperability also has a cost, which must be analyzed in relation to the benefits and opportunities it can generate to energy utilities.

How can interoperability be related to smart metering systems? There are many different ways. For example, in the home environment, one of the key devices that aim to provide interoperability is the in-home metering gateway. This gateway could interconnect different systems, including devices that can use different media, and translate different languages, formats, and concepts to form a common basis. It also aims to provide cooperation between different services and resources. This would enable synergies to be created by generating extra opportunities for both systems via the amplification of their related scope. It should also be a future-proofed device, to help the utility benefit from the existing infrastructure by providing potential future integration with other systems such as water meters.

Middleware plays its role, too. It provides interoperability between the smart metering systems and other system platforms inside the energy utility's corporate environment and with its utility partners. It equally provides interoperability between this system and its users.

Customer displays also play a fundamental role. These provide interoperability between smart metering systems and their customers. This interoperability is vital for customer satisfaction.

Finally, nowadays, much equipment is installed in homes without any interaction. For example, televisions do not talk to mobile phones, which do not talk to Internet modems. This offers an opportunity to in-home metering gateways to act as an integration platform for several in-home applications.

Technical aspects of interoperability have already been discussed in the technical architecture section of this chapter concerning middleware, in-home metering gateways, integration environments, and customer displays. The other key elements that should be considered are customers and users.

## Human interoperability challenges

Interoperability is a concept related to technology, is it not? If so, why are we talking about systems interacting with users and customers? In fact, interoperability between customers and technology is vital for these systems.

Let's reflect a little more on this. Maybe an energy utility sends you several letters each year. Do you even read them? If you do read them, do you read them in detail or just superficially? If you read them fully, do you truly understand them? Some of this correspondence could be for billing of the services they provide you. Do you fully understand this entire bill in detail? Do you really understand how to benefit from rates and services? Do you really know all your available services? Maybe you do—but do you think all the customers do? Maybe the utility is not interacting well with its customers. Maybe interoperability between their systems and their customers has failed. This may be due to several factors such as the communication language used, the difficulties of interoperating with energy meters, lack of customized services, and not listening to customer needs.

It is common to use the term "mandate." This term is sometimes not conducive to enabling interoperability between the customers and their utilities and systems. Sometimes utilities mandate services, solutions, and interfaces to be used by customers without really listening to their needs. Listening is not just asking for feedback through surveys, but understanding it and finding the most appropriate way of implementing feedback if feasible.

The concept of interoperability itself also applies to relationships between technologies and people. These two different entities should effectively communicate and cooperate despite their different language, environment, and conceptual models.

Can a system really be interoperable among all these different entities such as people, devices, and applications? Yes, if it has been developed for this purpose and the customers' needs and wants have been taken into account. First, customers must feel part of these systems from the start.

It is important that customers feel comfortable enough to voluntarily use the systems. They must feel they are owners of technology, not slaves to it. They must also realize the benefits they can receive. Selling services is a challenge, as it is difficult to achieve a level where customers realize the value of intangible products such as energy. The challenge is even bigger for public services, which may sometimes be seen as a free entitlement.

In order for utilities to readily interact, customers must be understood not as a group but as individuals. Of course, these individual needs may form groups of common services, but even so, there should be some personal design. Simple customization can sometimes make a big difference, such as simply changing the screen background and other personal settings in personal computers.

It is then important to consider other aspects such as culture, ethnics, customs, social aspects, and beliefs. This kind of interoperability is obtained in gradual steps. Some customers, for example, have an aversion to technology. Often, the first reaction to innovation is rejection. It is up to the energy utility to persuade their customers. From an interoperability point of view, it seems reasonable to use existing, mature, and recognized media to act as interactive points between customers and technology, such as televisions used as customer displays. Training and education campaigns are essential for customers to understand this new environment.

Services should also be offered in stages. Too many new services and too much innovation at the same time will only serve to create more problems for customers and utilities.

Users should also be considered, taking into account their differences, and interactions should be tailored to the audience. For example, as smart metering projects are essentially formed of multidisciplinary teams, interaction with engineers should be different from interaction with marketing people. Despite differences, they must also cooperate for the benefit of the utility.

Documenting processes also helps this challenge. Specific guides, norms, standards, and specifications must exist to ensure that the system meets all predefined requirements. Documentation is generally the basis of procurement, but may also apply to other procedures such as training and validation of product development.

## Norms, standards, and specifications

Standards and norms are usually guidance documents published by organizations such as the IEC (International Electrotechnical Commission), European Committee for Electrotechnical Standardization (CENELEC), or

American National Standards Institute (ANSI). Specifications are generally sets of requirements based on standards and norms used for different purposes, such as procurement processes and product development.

Standards and norms are very important for defining interoperability in smart metering systems. They are necessary in several different layers such as communication and system integration. They are also necessary to ensure interoperability between different entities in a market. An example is their applicability to industry data flows used by energy suppliers to manage change of suppliers by customers in competitive markets.

Standards and norms may cover many documents, defining several different points such as rules, protocols, and data exchange formats. They are used to ensure a common understanding for the different parties involved in a system and to solve disputes between them. They may involve diverse areas such as system communication, operational frameworks, and legal arrangements.

In general, standards may be classified as public or private. There are several aspects related to this choice. Open encryption algorithms, for example, may be seen as more advantageous because they provide freedom to utilities. On the other hand, they can create risks of breach of IT security, depending on the way they are written, their availability, and management. Choosing between public and private standards is difficult for utilities and dependant on their specific needs, therefore utilities must fully understand these standards. It is also recommended that these standards ensure that utilities are neutral about some components of the system, such as communication and meters.

In terms of smart metering interoperability, important aspects to be considered in specifications are protocols and data exchange formats. In relation to protocols, new development can create extra risks to the project, such as bugs and incompatibilities. As utilities are often risk-adverse companies, it seems sensible to adopt available, recognized, and mature options. An example of existing protocols is TCP/IP, which is a mature communication technology for computer-based applications and has been adapted to the metering environment. Although mature, it is important that future-proofed developments are included in the platform such as Internet Protocol version 6 (Ipv6). This is adequate to provide interoperability in all the different communication layers of smart metering systems. It is also a compatible approach for the different types of devices to be used, such as smart meters, smart grid devices, appliances, and gateways.

In relation to data exchange formats, there are several current formats available in different standards in the market. Sometimes they are not fully adequate for utility needs, although they can always be adapted via standards review or related specifications. It is important that competitors in the market work together in order to build benefits for the energy industry as a whole.

Several interoperability-related norms and standards are available in the market, and new ones are being or will be written. It is up to the utility to provide a full evaluation exercise in order to obtain the best benefits in terms of interoperability and other aspects for their systems.

# Conclusion

This chapter covers technical and nontechnical concepts relating to smart metering systems, such as technical architecture, communication media, data security, data processing, and interoperability challenges.

The objective was to provide some background on smart metering systems, their components, main challenges, risks, benefits, and some of the opportunities.

The importance of data security for smart metering systems now seems clear, based on the threats and risks present in its environments. You probably also realize that the choice of communication media and architectural aspects sometimes defines the limits of the projects as well as its related future-proofing capability and connected opportunities. Data processing was presented as a key aspect, but before defining its strategy, it is necessary to provide analysis in order to cover several criteria such as costs, benefits, risks, and opportunities. An efficient data processing system can provide several benefits to energy utilities, as good data are important to these systems. You probably now understand how a well-defined interoperability strategy can enable you to maximize or amplify the scope of your projects. It also helps utilities to obtain extra opportunities that might not be possible with smart metering systems as an isolated platform, but can be achieved when associated with others. Integration was also discussed as an essential element and how important customers and users are for interoperability, due to their frequent interaction with smart metering systems.

You should now realize how analysis of these individual smart metering components is crucial to the total success of these systems.

The next chapter provides even more details of the opportunities smart metering systems present. We discuss some of the innovative products being provided and discover why more and more utilities around the world are investing in mass rollouts of these systems.

# 3 International Analysis

We can now analyze the reasons why smart metering systems are being massively deployed around the world. The chapter starts with a general view on these deployments internationally as well as the main challenges. We then discuss different rates and payment methods. Finally, we explore the importance of these implementations for the different participants in the energy sector.

Motivation for deployments has already been briefly analyzed in the previous chapters, but this chapter provides much more detail. The main reasons for energy distributors and retailers to develop smart metering systems are to improve management of energy networks, enable differentiation in competitive market environments, and provide innovative services such as in-home automation.

For distributors specifically, in addition to enabling smart grid projects, smart metering systems themselves can provide energy networks with several benefits such as energy quality. They can also be sources of energy management through load switch devices and services such as load limiting and curtailment.

For retailers, new products can attract new customers and ensure the loyalty of existing customers. Examples are new rates, payment structures, energy-efficiency services, and meteorological updates via customer display units.

Other market participants, such as customers and energy market regulators, can profit from implementations. The environment is often a beneficiary of implementations through the rational use of energy. This may create a reduction of the total energy supplied by the utility and consequent reduction of $CO_2$ emissions. These levels generally vary according to the energy matrix of the country and period of consumption.

Despite these common motivations, there are many factors that affect a company's decision to massively implement these platforms. It may depend on the environment where implementations are located and specific conditions. For some, the main incentive can be revenue protection programs aiming at reduction of levels of nontechnical losses. For others, these rollouts are essential for the utility to survive in an increasingly competitive environment.

Challenges must be faced for these companies to obtain the benefits of smart metering. Risks must be managed. Because of the international trend of massive installations, it may be difficult to find qualified professionals to install these devices in the field. It could be equally difficult to efficiently control the quality of device production in the manufacturing chain. This happens when there is a huge increase in the production line and short time scales required by massive deployments. It may be necessary to engage new professionals to work in some industrial sites who do not have enough time to achieve required levels of skills and professional maturity to properly deal with these functions. Risks are similarly created due to impulsive action taken by manufacturers to win bids. These initiatives can result in a lack of quality in electronic components used in metering devices caused by the need for reduced production costs in order to remain profitable or at least to reduce related losses. Energy companies must be attentive to these risks.

In order to start our analysis, an international sampling of these projects follows. This study is based on my personal international experience during benchmarking exercises in several countries by chairing sessions, giving presentations, participating in collaboration efforts and workgroups, and visiting utilities, manufacturers, solution providers, and standards institutes around the world.

## International Benchmarking Exercise

Benchmarking exercises are essential to energy utilities to specify their smart metering projects. We are able to learn from feedback provided by other implementations. During this process, investigations of achievement may help companies to estimate opportunities, but error analysis is equally important. For example, this could alert enterprises to unknown problems up to this stage. Understanding these points is an important way to avoid wasting money and time.

The benefits to be achieved seem clear, but it is still necessary to dispel other misconceptions. There are still some organizations and managers who believe that the costs of these exercises are a waste of money, resources, and time. Of course, benchmarking exercises frequently require travel to meet with key people and to analyze the targeted systems and products working in the field. Travel budgets can be abused, though, and as with any other budget, they must be properly monitored and the results efficiently measured.

It is crucial to note that benchmarking exercises reflect ongoing scenarios. They should be updated often in order to identify changes and to confirm predefined assumptions. In addition, other interesting tools may assist in providing cost efficiencies. For example, Energy UK (formerly called the Energy Retail Association), which is composed of the main energy suppliers in the UK, maintains an interesting benchmarking tool on its website (http://www.energy-uk.org.uk/policy/smart-meters.html). This web page was created and is regularly updated under the auspices of the Supplier Requirements for Smart Metering (SRSM) Project.

The Energy UK benchmarking exercises provide information on spot market situations, behavior, decisions, and trends of projects. Utilities and other market participants continuously update these exercises. They are important sources of knowledge.

Based on its importance, the following sections provide an international view on smart metering implementations, arranged here by continent. The objective is to offer you a brief overview on deployments around the world. It will be presented at a high level, although for your own exercises, you will need detail at the lowest possible level, for example, precise information about specific areas of a utility. This is necessary because companies may differ even in the same political or geographical environment. The same could be true for internal areas.

## Africa

Africa is a very interesting place for benchmarking. Exploring its different countries and regions, we can find utilities using everything from basic electromechanical meters, manually managed, to very impressive smart metering platforms. There are offline vending stations for prepayment and sophisticated payment modes that use SMS (short message service) based on scratch card vouchers for credit purchase purposes.

Among other initiatives to develop smart metering platforms around the world, some important ones are currently happening in South Africa and are mainly focused on the electricity market. This country sometimes acts as a kind of incubator for the African continent in terms of emerging technologies. Successful metering systems and payment methods naturally spread to other countries on this continent once they have matured in the South African market. This does not mean that the other countries do not develop innovative technologies. Many interesting creative and high-tech initiatives are being seen all over Africa. However, even with the full independence of the other countries and their recognized skills for developing new solutions, South Africa is the focus of this benchmarking exercise because their technology at this time represents the smart metering trends in this continent.

**Updating existing platforms.** South Africa, apart from being a fascinating and beautiful country with regard to their people, culture, nature, and internationally renowned figures such as Nelson Mandela, is also a place with different social environments. These situations differ drastically from poor townships to rich regions in Cape Town.

Everything indicates that South Africa has the highest number of electric keypad prepayment meters in the world. Other countries in Africa, like Sudan, Senegal, Mozambique, and several others, have similarly adopted the same technology. As discussed in previous chapters, these meters can be installed either inside or outside customer premises, in special cabinets or on top of posts. For external meters, consumers are provided with a customer display unit to update their energy credit and monitor their consumption.

They also have very sophisticated and mature payment platforms for keypad prepayment meters. These systems were installed for different utilities and are probably the best implementation for this purpose in the world. These technologies use codes typed on the meter keypads in order to transmit commands to it. This code is locally called a token. When presented with this special token, these meters can also behave as credit meters. Meters are thus already able to switch from credit to debit mode, and vice versa. Their systems are also built to support this functionality. For these reasons and many others, they seem to be the natural precursors to smart metering systems.

Massive marketing campaigns have helped to establish prepayment meters as a standard in South Africa, and they are used by people of different social levels, regions, and beliefs. Endorsements by well-known personalities were used during these campaigns with the objective of disseminating the use of prepayment technology in the energy market, and good results seem to have been achieved.

These meters are embedded with a load switch device. Its management methodology is mature as millions of devices are installed in this country. Local solution providers and utilities have been implementing complex related tasks, such as meter reconnection and power limitation, for a long time.

Middleware deals with highly complex processes for management of credit and advanced rate structures, as well as data and asset management. There are many utilities using radio devices (sometimes a power line carrier, or PLC, is also adopted) to retrieve data from their electricity meters, to provide clock information for these meters, and to store advanced energy profile data in a dedicated memory (fig. 3–1). This energy profile provides curves for customer consumption, credit history, quality of supply such as voltage, and several logs such as electronic counters for terminal cover removal. Status and logs are handled by available public standards.

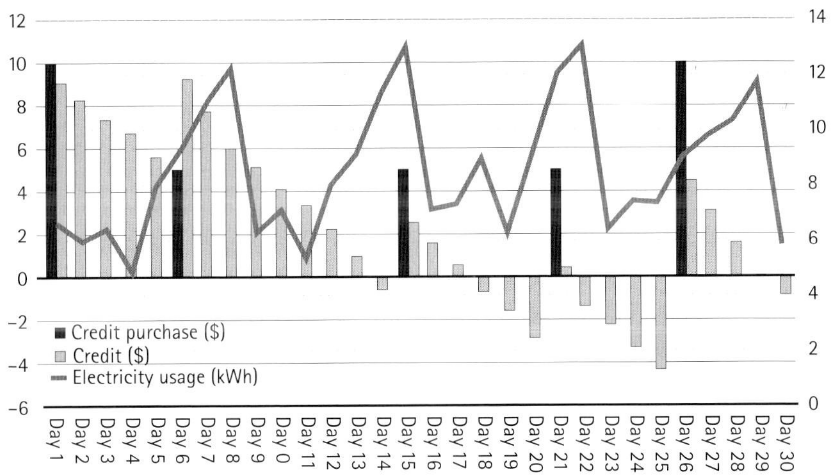


Fig. 3–1. A load profile combining credit and register history

The next step is the addition of data concentrators to manage these PLC and radio frequency (RF) systems. This assists these systems to become AMM (automatic meter management) platforms.

The majority of these meters meet the STS/IEC (standard transfer specification/International Electrotechnical Commission) standards, which define, among other things, management rules and encryption algorithms to create the data exchange tokens. These standards provide interoperability between meters produced by different manufacturers as well as middleware platforms produced by different solution providers.

Thus, they give utilities the freedom to use interoperable solutions from many meter manufacturers and solution providers in their installed base.

Based on this, these systems are likely to be converted to smart metering platforms in the future. The main difference is that the tokens, at this stage, are automatically sent to the meters via communication media instead of being transmitted by typing codes on keypads. Utilities are also able to retrieve data from the meters without the need of sending technicians to the field with radio-equipped or PLC-based handheld units. These systems need only to be provided with a two-way wide area network (WAN) technology to have some new features implemented and adapted to this new environment. The STS/IEC standards also meet smart metering requirements.

In practical terms, there is much work to be done because of some specific issues related to smart metering implementations. However, part of the way has already been paved. For example, STS organizations have been working for a while in this direction because local utilities, manufacturers, solution providers, and other STS members understood both the international trend as well as other market needs such as generation capability issues.

Several existing features of these standards can add value to smart metering systems. An example is the use of tokens for credit management and respective processing algorithms such as token validation in the meter and respective logs. During this validation process, the code may be accepted, corrected, or rejected by the meter. For example, the code may be rejected due to an incorrect format, its validity is expired or it is already used, it is the incorrect meter, or the meter is programmed with an incorrect rate type and rate. This advanced processing on the meter side may add many advantages to the systems. In the smart metering world, tokens may continue to be given to customers in order to be used as a contingency option. They could be adopted in the event of a communication failure. In this case, the customer could type the code into the meter or in a customer display unit. If the code was already sent by the middleware, it would be rejected; otherwise, the meter would process it. Customers would then be provided with proper feedback and credit updates even during communication failures.

As communication failures are sometimes unpredictable events, contingency action is important to ensure customer satisfaction. In order to illustrate this need, imagine an extreme scenario where customers are on the limit of their emergency credit. This credit is enough only to continue to supply their premises with energy for 5 minutes before self-disconnection, and the customer purchases extra energy credit at the last moment. At this stage, a communication failure can happen between the middleware and the meters. If its duration is more than 5 minutes, the customer can be disconnected because of this lack of credit update. This could be a high risk for utilities. If at the moment of credit purchase customers were provided with a token and instructed on its use, this situation could be avoided or at least minimized by the customers themselves by typing their codes on the meter's keypad.

A physical keypad is a necessity. Touch-screen interfaces or even simple arrangements of just two buttons—the option present in the majority of meters—could be used. Thus, this functionality can be implemented for utilities at very low cost.

**Reasons to deploy smart metering systems in South Africa and proposed solutions.** Natural evolution of the existing platforms and standards are not the only reason that encourages South Africa and others countries in Africa toward smart metering. Recently, the country has experienced a number of blackout events. The main problem appears to be limited generation capability, insufficient to meet the increasing demand. New investments in generation have been made, but these initiatives generally take a long time to be fully implemented. Smart metering systems can be a potential solution to deal with this problem in the short and medium terms.

Among other drivers, improvements in energy efficiency management, reduction of energy consumption especially during peak times, load curtailment and limitation management, improvements in the quality of electricity supply, demand-side management, demand response rates, and grid management are the main reasons for massively deploying smart metering systems in this market. Recent announcements demonstrate that authorities are deciding to move in this direction.

New laws and market rules have been enacted requiring energy-efficiency action in South Africa to deal with a potential energy crisis. These initiatives strongly recommend the use of smart metering platforms that can optimize energy consumption, for example, through heating controllers and pumps for swimming pools (fig. 3–2).

Many market participants demonstrate a desire to manage appliances in homes. Customers, for example, would be able to choose what appliance to turn on and off during different periods of the day. This can be configured according to sets of automatic rules that define priorities among the different devices. They can be related to rates and other products. Switching events can also be manually executed if desired by the customers.

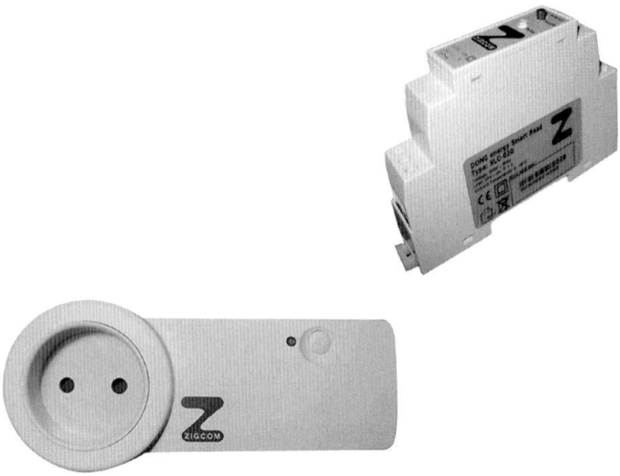


Fig. 3–2. Relays for demand-side management deployed in South Africa. (Photo by Dorthe Gaardbo-Pedersen; courtesy of Develco.)

Advisory systems can be implemented to assist customers with energy-efficient actions. For example, with the agreement of customers and subject to legal requirements, these platforms could inform customers when they were using more electricity than other customers with similar habits and premises. Action can then be proposed according to intelligent analysis of their load profiles.

The main energy utilities in the South African market have already announced their intention of installing millions of smart meters in the next few years. Everything indicates short- and medium-term trends of massive deployments in the country, at least in the electricity market. This will soon be disseminated to other countries on the continent.

## North and South America

North America, together with Europe, forms part of the largest deployment of smart metering systems in the world. This is not the only part of the continent where these rollouts have been taking place. At lower scale levels, but with similar importance, these systems have also been installed in Central and South America. In general, smart metering systems in North America are being developed for multifuels. They are driven by different factors depending on specific needs.

**United States and Canada.** The United States and Canada are usually recognized for their high-tech initiatives and strong action on research and development in technology. This trend is reflected in their smart metering implementations. Although other countries in this region, such as Mexico, have also been developing smart metering systems, preference is given here to describing the ones from the United States and Canada in more detail due to their large deployments at this stage.

Overall, initiatives in both countries focus on energy management with the objective of consumption reduction, mainly related to peak time. This creates action in several areas like energy efficiency, network management, demand response rates, demand-side management, and change of customer behavior. The main drivers are optimization of daily load curves and balancing demand to the available production capability. Strategies are several and vary from utility to utility. For example, there are utilities offering advanced platforms where heating controllers are integrated with in-home metering gateways and smart rates.

The United States and Canada are responsible for the biggest initiatives on the continent, involving tens of millions of meter points. These rollouts have been facilitated thanks to their experience with existing AMR (automatic meter reading) systems. These platforms are now mature and were installed over many decades. They are now being updated with more sophisticated systems such as AMI (advanced metering infrastructure).

In the United States, for example, low-power RF systems have been massively implemented for many years to collect metering data remotely and control meters. This level of maturity provides the potential advantage of upgrading these platforms to smart metering systems. Several functions related to metering data management are already solidly implemented. Integrations with other utility systems are also in place in several cases, as well as advanced rate schemes. This provides opportunities to amplify the scope of smart metering projects based on acquired experience with these AMR platforms.

Note that decisions involving communication media are not always connected to the technology itself. For example, due to network topology, PLC systems are often more cost effective for utilities in Europe than in the United States. In Europe, power transformers often supply hundreds of customers, while in the United States, they only supply tens of customers or even only few. This is a result of different aspects such as strategic decisions when building electrical networks and distribution of houses around cities. As PLC signals are unlikely to be sent from the secondary (low voltage) to the primary side (medium voltage) of distribution transformers without using auxiliary devices, the cost per node is higher in the United States for this technology, taking into account that costs of these devices and data concentrators should be shared per node. For this reason, RF technologies have been widely used in the United States and PLC in Europe. Of course, there are some exceptions; for example, there are utilities using narrowband and broadband PLC in the United States as part of their smart metering strategy, and vice versa for Europe in terms of RF technologies.

In 2005 the U.S. Congress passed the Energy Policy Act, which put in place federal action to enable demand response management services. Strategies toward smart metering systems seem to be the logical way to go, although it is up to individual states to implement them. In 2009, U.S. President Barack Obama, via the American Reinvestment Recovery Act, announced a public investment of $3.4 billion in smart grids (total private–public investment estimated at $9.3 billion). States and energy utilities have been making different decisions related to technologies and implementation strategies.

Some utilities are investing in very sophisticated demand response rates, others are more focused on the load management, and others are making good use of the opportunity to implement broadband technologies that enable the ability to provide innovative services such as broadband Internet access and voice over Internet Protocol (VoIP) to their customers.

Another common objective is the reduction of operational costs and reduction in the price of energy purchased by utilities from generators. For example, costs are generally higher during peak times, and reduction of peak load can contribute to reduction of the total energy price.

In terms of supplying remote areas under extreme conditions, Alaska is a good example. Alaska is the largest state by area of the United States. Although there are difficulties in supplying energy to this region, due to its large geographical area and extreme cold weather conditions, smart metering systems have been successfully deployed by the Alaska Village Electric Cooperative (AVEC). AVEC, according to its website, is a non-profit electric utility, owned by the people. It serves 53 villages throughout interior and western Alaska. In fact, the AVEC service area is the largest of any electric cooperative in the world. This is a clear example of how robust a smart metering solution can be and how large the range of benefits can be.

In Canada, a directive requires the installation of advanced meters. In Ontario, authorities are actively participating in the design of the smart metering platforms and deployment strategies. Beyond these similarities with the United States, Canada has been using time-of-use prices to encourage changes in customer behavior. These prices take into account when and how much electricity has been consumed by customers. Thus, prices vary during the day, based on production costs. They are equally dependent on seasons and specific periods when energy is consumed. Different rate periods have been defined such as "on-peak" when energy is more expensive; "mid-peak," less expensive than peak; and "off-peak," the most cost effective of all. This scheme can be very advantageous for consumers.

There are good opportunities for utilities to reconcile the cost of energy and its price. Energy is purchased by utilities at several different prices during the day. Basic rates are unlikely to provide the required reconciliation between loss and beneficial margins to utilities, depending on the time energy is used. On average, this margin tends to be positive, but schemes need to be optimal for these companies and their customers. Government is also engaged with renewable energy action, and demand response management is compatible with this policy.

**Central and South America.** Central and South America have similar needs as North America, yet sometimes associated with social and economic problems. Despite natural beauty, culture, diverse people, and several other positive aspects of different environments in parts of the continent, some utilities suffer with high levels of nontechnical losses, late payment, and nonrecoverable payments.

Energy theft is a challenge faced by many utilities in South America. Revenue protection action has been taken to try to counteract this, but it is mostly ineffective because of other aspects such as technical rather than social activity, difficulty in targeting the losses, and a lack of necessary technology. There is also a trend in some countries of moving meters to outside a house or reconfiguration of network topologies, among other actions, to facilitate management of revenue protection.

The same challenges apply to debt management. Prepayment meters have been installed in several utilities in these regions to try to mitigate these problems. Some actions are successful, although the cost to maintain these technologies is sometimes prohibitive.

Smart metering systems can provide effective solutions to deal with these problems, thanks to the level of technology and business intelligence tools. Advanced pay as you go (PAYG) functions can help customers to meet their budgeting needs (see the section on PAYG in this chapter for more information). Energy displays encourage customers to gain a better understanding of their related demand. Benchmarking exercises tend to encourage a reduction of consumption when these devices are installed.

Taking into account the levels of associated financial loss, these rollouts can pay for themselves. It is also important that these projects are part of a social strategy to ensure that the envisaged results are obtained in all targeted areas. Although the benefits in terms of business cases are high, projects should not be restricted to revenue protection action. Customer satisfaction must always be the priority. Other benefits should also be considered for all of the market participants, such as energy efficiency and carbon reduction initiatives.

Many smart metering project initiatives can be found for customers with high energy demands (fig. 3–3). They use online intelligent systems and algorithms to deal with revenue protection and add value for customers such as energy quality improvements, rate advice, new web-based displays, and demand-side management. Among other initiatives, Light, one of the main electricity utilities in Brazil, has built a smart metering monitoring center that can manage hundreds of thousands of its customers online.

Fig. 3–3. A metering gateway device in Brazil for customers with high energy demand. (Photo courtesy of CAS Tecnologia.)

There are some massive implementations in these regions involving remote management of metering systems for residential customers. However, they do not necessarily bring together all the opportunities and functionalities that smart metering platforms can offer due to different aspects such as lack of specific rules and regulations (under preparation) for some countries. Regarding joint smart grid programs, there is still a long way to go for many of the participants of this market, particularly in the residential market. There are a few notable exceptions, for example, Light and Cemig, together the largest energy utilities in Brazil, are responsible for providing energy for more than 10 million customers in Rio de Janeiro and Minas Gerais. In September 2010, they started one of the largest and comprehensive R&D smart grid programs in the world. In general, though, many challenges and limitations must still be overcome to enable these deployments. Specific regulation and directives are equally necessary to set the terms for these potential rollouts. Because of this, most activity in this area is still in the initial phases and implemented in low-volume programs such as pilots, research and development, metrological evaluation, and regulatory analysis. For example, the regulatory authority in Brazil, Agência Nacional de Energia Eletrica (ANEEL), has organized a public forum on the installation of smart meters (potentially 56 million customers) and has issued a resolution that contains the minimum functional requirements for smart meters. This is an important step toward smart metering systems.

## Asia

This continent contains the largest population of all and thus the biggest potential volume opportunities for smart metering deployments. This region is equally recognized for high-productivity capability in the technology sector. Millions of devices are produced every month in this region. Asia is also famous for research and development of high-complexity technologies.

Due to their capacity and the low cost of metering devices in production, a large share of the smart meters to be installed around the world will probably come from this market. During my visit to this continent, I saw promising features of productivity, technology, and research and development.

Despite this promising environment, the extent of nontechnical losses, in some Asian countries and regions, is the highest in the world. Levels of tamper sophistication are also impressive. There are similar problems in debt management. In order to address debt, utilities have used revenue protection and prepayment schemes, but still suffer problems similar to those in South America.

Losses are one potential driver for smart metering deployments in these regions but not the only one. Energy efficiency is another. The region is already one of the most advanced in this aspect in the world; for example, the advances made in Japan are impressive. Many advances have taken place in this area of the world over several years. Devices such as refrigerators, air conditioners, and washing machines contain intelligence for optimizing self-energy management and being interoperable with other devices and systems in the home. Similar work has been developing for light bulbs and water heating.

Smart metering systems give an opportunity for these platforms to be integrated with the metering environment. They can provide intelligent rates and other products to be offered to consumers by energy utilities and/or their partners.

Energy consumption management is another reason for implementation of smart meters. Despite the unbelievable investments and achievements in terms of energy production that enable new power plants to be built in this market every year, generation capability is unlikely to support the growth of their population. Because of this, action on the rational use of energy and reduction of peak loads are welcome and needed.

There are opportunities for these deployments to integrate existing and new smart displays, home automation devices, and advanced gateways with high interoperability levels. For example, innovative technology research has been done on this continent such as data transmission using human interaction. This method can safely enable communication with different devices, for example, using a handheld computer as a display unit. Data could be transferred from a home gateway to this device (assuming that it is somewhere near the customer, inside a pocket, for example) when the customer touches sensors located in the home, such as on entry doorknobs. This method presents several opportunities to many market participants, including energy utilities.

In addition, there are various types of intelligent metering technology, such as meters with embedded intelligent algorithms and smart measurement circuits (such as those used in India), which are able to detect and compensate for several types of tampering. This means that even during tampering this meter can maintain its measurement accuracy. This feature, with a smart metering system, can add implementation benefits for revenue protection.

Some utilities have already announced potential deployments of hundreds of millions of smart metering systems over the next few years, using different technologies. Although the majority of utilities still seem to be in a phase of research and development and low-volume pilots in the field, there are some smart metering initiatives in this continent. For example, in Korea hundreds of thousands of smart meters based on code division multiple access (CDMA) technology have been deployed at industrial sites by local utilities. China has also announced the intention of replacing hundreds of millions of meters in the coming years.

Original equipment manufacturer (OEM) providers located in this continent are also working with some key international smart meter manufacturers in order to design and produce the potential millions of devices that will soon be needed in the world. Proper procedures must be in place to ensure the international required levels of demand are achieved, such as safety, socially responsible policies, metrology, and quality. This strategy seems to be an international trend, illustrated by the fact that Asian manufacturers are currently responsible for providing tens of millions of advanced metering devices used for the biggest AMM and AMI deployments in the world.

## Europe

In terms of the number of deployed solutions and initiatives, countries involved, and diversities of deployments, there is no doubt that Europe is currently one of the biggest environments worldwide for smart metering systems.

The drivers to deploy smart metering platforms differ from country to country, according to their specific needs and market organization. In some regions, for example, these platforms have been deployed to enable competition in energy markets. Sometimes these rollouts are executed as part of the process of opening the market for competition. For existing competitive markets, these deployments have been used to increase the package of products offered to customers in favor of more aggressive churn rates. Thus, it is clear that these platforms are crucial to enable differentiation among energy companies in order to increase their market share and to ensure customer retention.

Revenue protection is another driver for some deployments in Europe. Even if nontechnical loss levels in this continent are not as high as in some regions of the world, controlling loss can still offer important financial returns to business cases.

In other regions, smart metering implementations seem to be in response to authorities' requests, white papers, or directives such as established obligations to reduce estimated bills.

The main reasons for these initiatives can differ but, in general, they share some common drivers, for example, reduction of total consumption (mainly during the peak times), quality of energy supply, energy management, carbon emission targets (often based on the Kyoto Protocol), change of customer behavior in favor of their energy efficiency, social responsibility goals, and encouraging the deployment of renewable sources of microgeneration.

Tens of millions of multifuel meters have been installed in Europe (fig. 3–4). However, there is still a potential to implement even more in the near future by different utilities. In countries like France, energy distributor companies are responsible for metering; in other countries, such as the United Kingdom, energy suppliers handle the metering. These different business ownerships are due to specific market organization, policies, laws, strategy, and regulation.

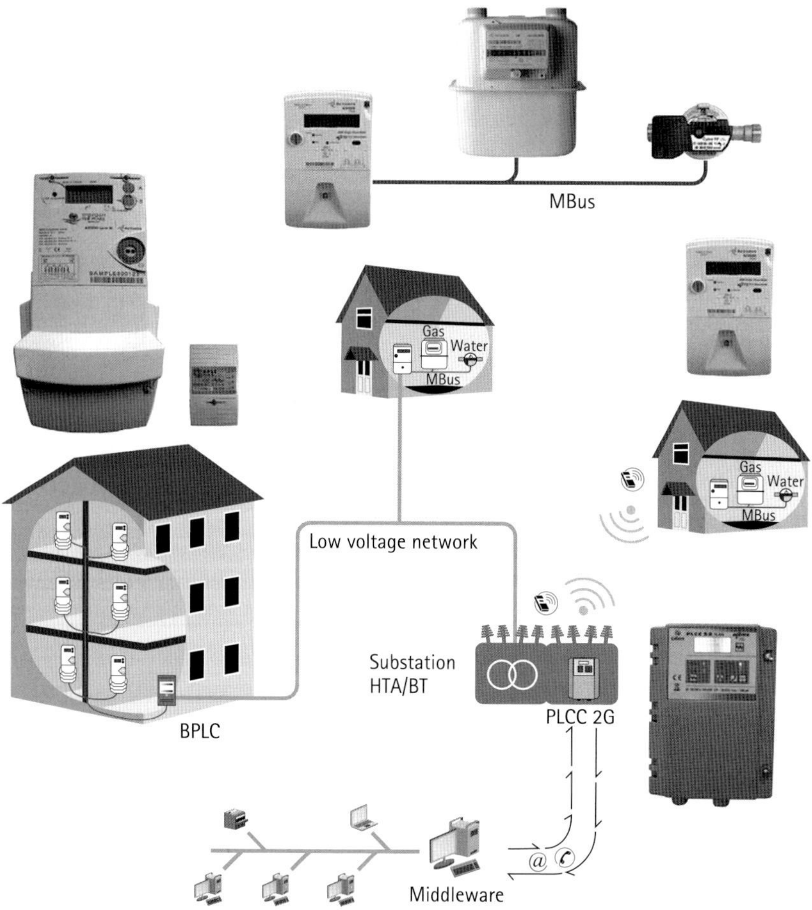


Fig. 3–4. An example of a smart metering system deployed in Europe. (Illustration courtesy of Actaris.)

**A vision per country.** The first massive implementation in the energy sector in Europe was in Italy by ENEL, Italy's largest power company. Around 30 million electricity meters are in this AMM deployment in an investment evaluated in billions of euros. ENEL claims to have obtained positive results in this initiative.

Markets have been organizing in different ways to execute these rollouts, depending on local strategies and market organization. In some countries, such as in the United Kingdom, suppliers are joining forces on a common project to avoid the need to replace meters when customers change energy suppliers. A common specification was developed by Energy UK through the SRSM project. There are similar developments of common standards to ensure smart metering interoperability in other countries, such as the Netherlands.

Scandinavia is currently responsible for a large share of the massive rollouts in Europe. It has been installing millions of smart meters in the field. Different utilities in this region have adopted their own strategies of deployment, making this probably the most hybrid deployment on the continent. Some utilities, for example, invested in PLC architectures associated with low-power RF solutions. Other energy companies are deploying RF mesh solutions or Global System for Mobile (GSM)–based ones. In Sweden and Finland, massive deployments are at an advanced stage. Highlights in this region are the massive rollouts by Vattenfall and Gothenburg Energi, two of the main local energy companies.

Other utilities have decided to adopt a rollout strategy based on targeting specific segments of customers instead of a massive deployment. An example of this scheme can be currently seen in Denmark. DONG Energy, one of the main local energy utilities, is responsible for an advanced targeted rollout that offers home automation functions to their customers, such as remotely commanded smart plugs and security alarms, through their smart metering platform. However, massive deployments appear to be a trend in Europe. Massive rollouts can offer customization levels per customer segment, but they also enable all the customers of an energy utility to make use of the benefits offered by smart metering systems.

In France, ERDF, the energy distributor part of EDF Group, has been deploying a large pilot project of smart metering called Linky, whose objective is the replacement of the 34 million residential meters in France. The Linky project is based on a three-layer architecture—IT, data concentrators, and meters—and is focused on the metering goals of ERDF. The meter contains interfaces that enable the different energy suppliers to offer different types of services (information, demand-side management, etc.).

ERDF is also developing a third-generation PLC technology that will allow a tenfold increase in communication bandwidth between data concentrators and meters, compared to the currently available PLC technologies. This technology, based on OFDM (orthogonal frequency division multiplex), will ensure that the system is future-proof.

Spain has also worked in this domain. Iberdrola, one of the largest energy companies of Spain, has is developing a technology called PRIME that is based on OFDM.

In Germany, some utilities decide to opt for what could be classified as an aggressive technological strategy, in the best sense of the word. For example, Yello Strom, part of the EDF Group, is installing meters based on Internet data exchange with a very high-tech design. The metering industry is definitely experiencing a revolution. This system has been providing home automation, such as automated light bulbs, and offering online load profiles via a web-based platform. This initiative also includes future integration with advanced high-tech displays such as modern mobile phones (fig. 3–5) and Chumby devices (fig. 3–6). Chumby is a compact Wi-Fi device that displays useful and entertaining information from the web.

Other countries in Europe have been developing smart metering technologies in pilots or medium-volume implementations of tens of thousands of meter points. The UK is also responsible for technology pilots to prepare for a potential massive rollout. EDF Energy, part of EDF Group, in association with the local government, universities, innovation companies, and other energy suppliers, is part of this initiative. Their pilot projects include advanced interaction with customers using complex home automation systems, such as heating controllers integrated with smart metering gateways. It also offers interactive channels such as television, customized customer display units, web pages, advanced reports, and smart payment platforms such as PAYG.

Finally, many other utilities in Europe, such as Endesa, a leader utility in Spain and Eletricidade de Portugal (EDP), the biggest electricity utility in Portugal, are looking toward smart metering.

**Common features.** Meters, in general, should be able to measure incoming and outgoing energy in order to comply with renewable energy strategies such as decentralized microgeneration. Meters should also enable submeter arrangements for measuring total micro-generation gains. Reactive energy measurement is also an option in Europe.

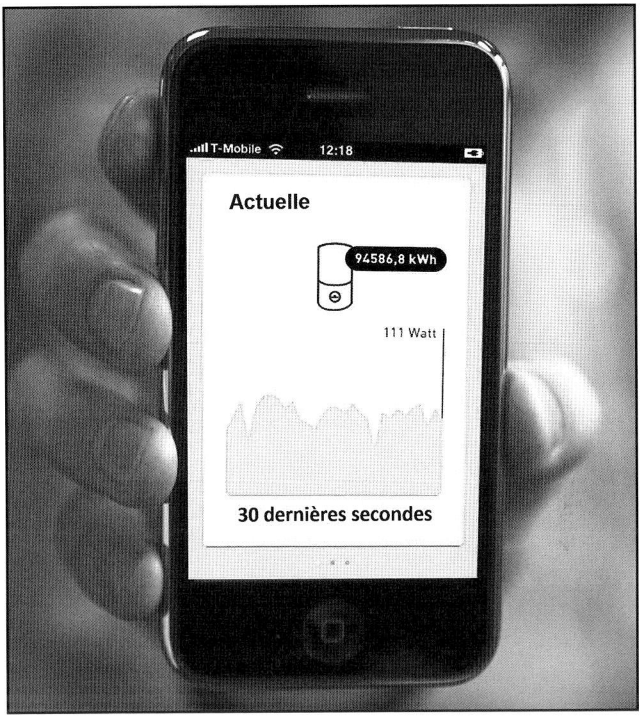


Fig. 3–5. A smart metering system integrated with a modern customer displays on a mobile smart phone. (Photo courtesy of Yello Strom.)

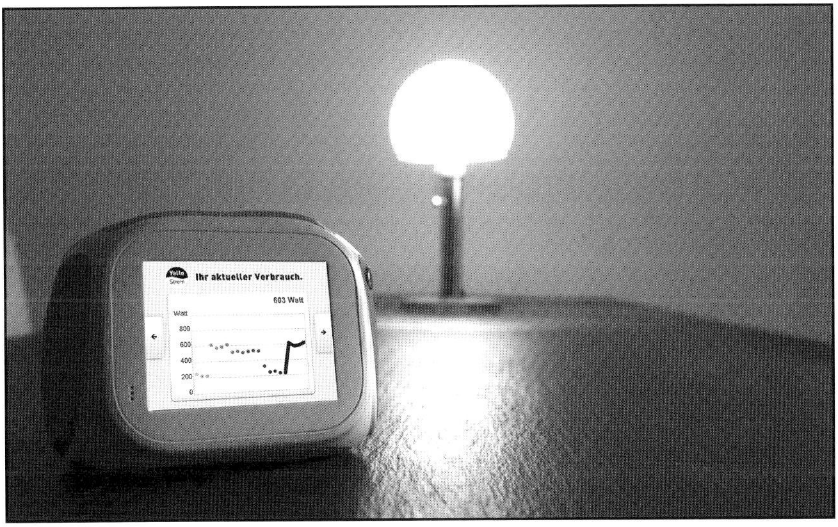


Fig. 3–6. An example of a smart metering system integrated with modern customer displays on a Chumby device. (Photo courtesy of Yello Strom.)

There are different types of WAN and local area network (LAN) communication technologies currently used for these implementations. Examples are PLC (fig. 3–7), mesh radio, and GSM for WAN. In terms of home area network (HAN), low-power RF mesh-based seems to be one of the most used (fig. 3–8).

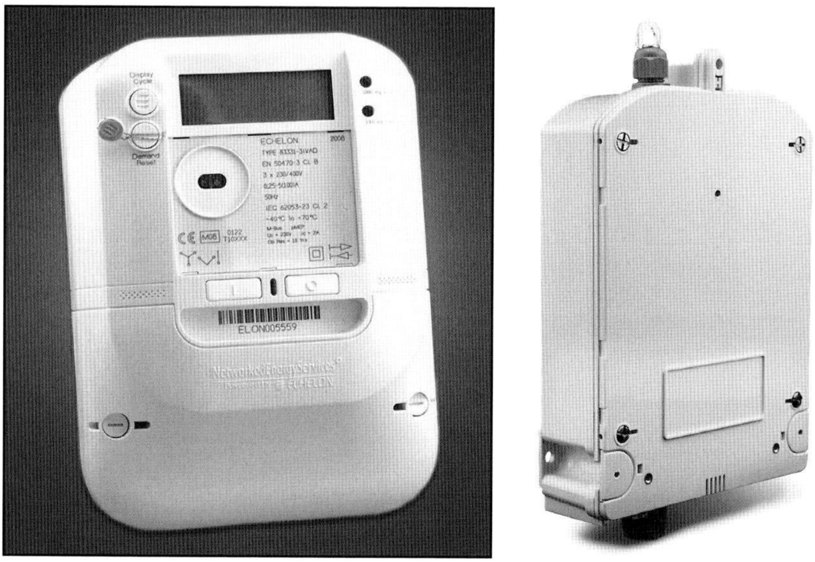


Fig. 3–7. A smart meter and data concentrator deployed in Europe. (Photo courtesy of Echelon Corporation.)

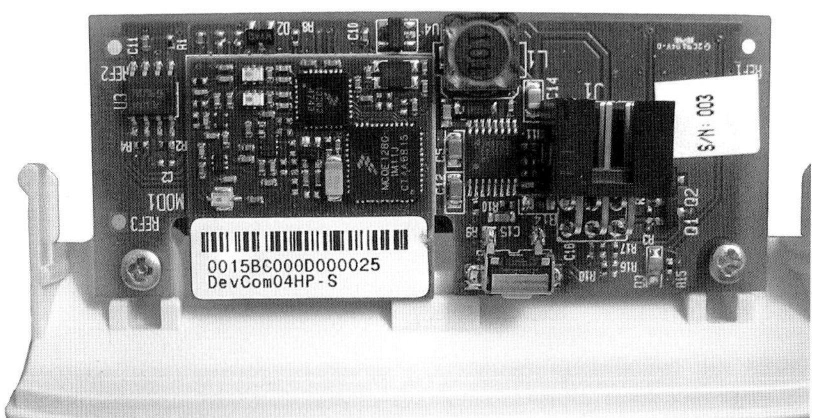


Fig. 3–8. A PLC-RF bridge device deployed in Europe. (Photo by Dorthe Gaardbo-Pedersen; courtesy of Develco.)

Thanks to HAN capability, dual-fuel arrangements are commonly deployed by utilities that provide both gas and electricity to their customers. Displays are used for both fuels to provide customers with a total energy view. Sometimes monetary conversions and financial forecasts are also supported.

Some utilities are making good use of smart metering platforms to provide interoperability with in-home devices such as smart plugs and movement sensors. Services such as security monitoring, demand-side management, and reductions of standby consumption for some types of equipment have been piggybacked with in-home metering gateways.

Displays appear to be a trend in some countries. They vary from customized basic devices to complex devices and widget interfaces. There is a high degree of confidence in some countries that the use of display devices is essential to changing customer behavior. Thus, they are essential to utilities, government, regulators, and other market participants in achieving their targets, such as reduction of energy consumption and carbon emissions.

Smart thermostats, heating controller systems, and water heating storage have been part of several pilots involving better management of heating systems rates. Another trend is load shifting, which may also be the future for intelligent appliances.

Some markets, such as the United Kingdom, have a high share of prepayment meters. They were generally seen as having high costs of maintenance and management. However, the ability of the meters to switch from credit to debit mode reduces these costs significantly and offers even more efficient payment methods. Thus, this enables more opportunities to utilities and customers in competitive markets.

Data transmission and technical architecture vary from implementation to implementation in Europe. Some utilities are investing in online infrastructure where customers are provided with local updates every few seconds on their display units and every quarter of an hour on a web page interface. There are also other applications moving toward load profile updates only once a day and restricting online services for high-priority messages such as alarms and rate signals.

The authorities, in general, seem to be actively participating in these processes. In some countries, working groups were created to discuss the subject, and investments have been made by some governments that will partly or fully finance deployments and pilot initiatives.

As you can see, the strategies to implement smart metering systems in Europe vary according to local requirements, but they are definitely not a trend but a reality, given that tens of millions of devices have been installed in the field and are being managed by complex IT platforms. They are additionally integrated with advanced in-home platforms, which expand the opportunities for their customers. However, this is just the beginning, and there is still much more to come. Tens of millions more meter points are due to be implemented in the next few years.

## Oceania

Oceania is a continent composed of mainly islands. This kind of territorial arrangement provided an ideal opportunity for smart metering systems. Because some of these islands have difficult access and low population, utility companies can see efficiencies in managing their measurement devices remotely. The main initiatives in this continent for smart metering deployments are located in Australia and New Zealand. Recent announcements by some local utilities estimate the potential deployment of millions of smart meters in both countries in the next few years.

Major drivers for these projects are reduction of peak energy consumption, energy efficiency, promoting competition, improving price signals, enabling demand-side management, and offering other innovative services to the energy sector.

Improvements in the price signal mechanisms seems essential to their deployments because it can enable customers to manage their demand and cost of energy better through the day. Customers can receive price signals with different user-friendly graphical formats, such as color schemes. These formats help customers to understand their consumption and adapt their behavior to these innovative rates.

Optionally, an automatic load-shifting service can be in place. For example, a water heating storage system can be controlled via HAN by the in-home metering gateway. It could be controlled different times a day, depending on customer needs and on the price of energy. The tank can conserve hot water for many hours thanks to its insulation level. In case of an emergency need, the user can manually control the system. Similar functionality can be implemented for appliances such as washing machines. The machine can be loaded any time of the day, but it would be automatically controlled by the in-home metering gateway when the rate is more economical to the customer. Of course, full interoperability with the device is necessary for the smart metering system to receive feedback on the periods when demand-side management could be executed. For example, laundry cycles should not be interrupted in some stages.

Another important feature provided in the project is the capability to measure import and export energy. This enables compatibility within the deployment for distributed renewable microgeneration sources. Customers may receive different financial incentives depending on the energy source used (e.g., photovoltaic, wind-based, and others) on renewable certification programs. Given that one metering element should apply to each source, submeters are necessary. In the UK, instead of using submeters for measuring renewable energy (as in Europe), for single-phase installations, multi-element smart meters have been used. This eliminates the use of extra space in customer premises for these applications. They have been developed and are wired so that the first element is dedicated to measuring main consumption (net between import and export), and the others are dedicated to measuring locally generated renewable energy. The same principle can be used for plug-in electrical and hybrid vehicles, if desirable for the utility or necessary, depending on the rate structure and incentives involved in the implementation.

Highlights for these projects are advanced displays (fig. 3–9). Beyond rate price signals, they are able to reflect instantaneous consumption levels, represent historical consumption in graphical forms, provide monetary conversion, and offer several other features to help customers efficiently manage their energy budgets.

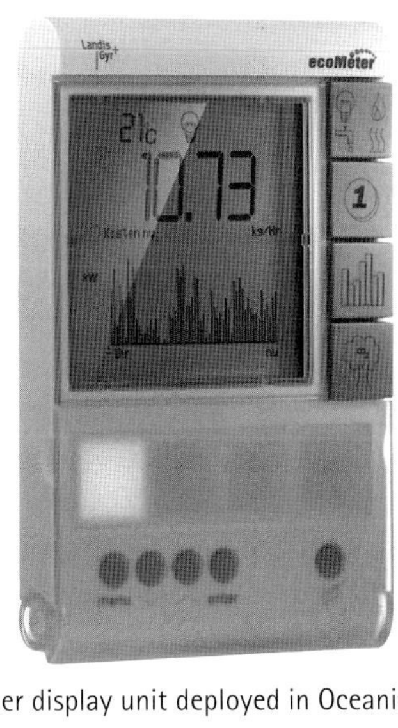


Fig. 3–9. A smart customer display unit deployed in Oceania. (Photo courtesy of Landis+Gyr.)

Several pilot programs have been conducted to enable local utilities and authorities to understand customer reaction to different technologies and services. These initiatives were split into different phases. During this process, the benefits to the different participants of the local energy market were measured. Deployments take into account this feedback in order to achieve the required level of customer satisfaction and other market targets.

# International Drivers and Opportunities

During an efficient benchmarking exercise, it is common to identify the potential opportunities and challenges a utility can deal with, based on experiences with similar projects. Even if projects vary depending on the particular implementation aims, synergies are likely to be identified.

In this section discuss some potential opportunities and drivers associated with the deployment of focused technology. We discuss, for example, some natural factors that almost require pilots and deployment of smart metering systems. Some can be used in business cases and technical viability studies and analysis. These points are discussed at a high level only. Investigating these topics in details and providing continuous updates should be a goal of the market participants.

## Natural drivers to deploy smart metering systems

Building a smart metering project is sometimes very arduous because of the analysis necessary to ensure the success of the project. Environmental analysis is one of the most important to identify opportunities and challenges for smart metering projects. It is essential to define the entities around this project and also the risks, challenges, drivers, opportunities, and other aspects that can influence the deployment decision.

Incentives to deploy these systems can emerge from different entities and stakeholders, including meter manufacturers, universities, technology research institutes, governments, customers, and others.

**Local authorities.** Local authorities have been studying smart metering in many countries. The energy supply service affects the population at several levels, such as comfort, satisfaction, life, and health. It can also affect carbon emissions, changes in the environment due to the need to build new power stations, and many other factors related to the public interest. However, other aspects can be considered such as development of direct and indirect employment; avoiding natural market trends; global warming; and improvements in diverse economic sectors, renewable energy, and innovation.

For these reasons, authorities sometimes move beyond their standard market responsibility to legislate and regulate energy utilities in order to provide incentives and assistance to them to develop these projects. This active participation is important to ensure that these projects meet all the needs and interests of all the participants of the energy market. Assistance varies from providing specialized resources to organizing working groups to partially or fully sponsoring research and development projects, field trials, and deployments.

Of course, authorities also require targets to be achieved and continuous audits to ensure that the investment is being properly applied for the population, the environment, and other applicable areas. This potential driver can help utilities to develop these technologies properly and to ensure that their social and environment commitments are achieved. It is also important to mention that some authorities in the world are studying methods that could actively help to reduce carbon emissions resulting from energy use. Although energy utilities have been trying to move demand to times when energy is cheaper, the network is less loaded, and less pollution is generated, the budgetary aspects sometimes do not incentivize customers because the financial values are not large enough. In several countries, taxes are a large part of the rate, irrespective of the period energy is used. To deal with this challenge, studies have been performed to apply different values to taxes depending on the time electricity is consumed. This may be in important driver for smart metering systems and may produce benefits for all the market participants and the environment.

**Mass technology effect.** Smart metering systems are a reality in some continents, not just a trend. Several rollouts of these technologies are underway around the world with tens of millions of meter points. It is not an exaggeration to say that hundreds of millions of points will be implemented internationally over the next 10 years.

As with any other technology, the mass technology effect creates a reduction of costs. These reductions have already been observed in the metering market over the past few years, and costs will continue to go down even more aggressively. This trend applies not only to all the devices in the metering system infrastructure but also to other related components, such as IT software licenses and telecom costs.

Some markets are also mobilizing themselves to increase this effect even more. Utilities are organizing grouped offerings based on common minimum specifications. Customizations to enable differentiation can be achieved at the in-home metering gateway level. New applications can be built by utilities or third parties to be downloaded onto this gateway, because of the flexibility of this technology.

**Partnerships.** Participants such as telecom providers, customers, and universities, are very interested in this market. Based on common interests, the infrastructure can be developed at very low cost or even where the benefits outweigh the costs. For example, if a broadband PLC system is used, a third party can offer Internet service to utility customers. Common technology research is another way to save money as a result of potential common institutional interest. Another point is the ability to provide connectivity to smart metering systems by piggybacking Internet access through the customer's existing broadband modem. This could bring a significant reduction in communication costs for utilities.

Universities have a special interest in smart metering. Millions of different devices, operating in complex environments and interacting in the field, face challenges such as interoperability, data security, and data management. Research and development at the university level could be a unique opportunity for doctoral candidates, and the universities could reap financial rewards in the form of product development, patent publications with associated royalties, and marketing opportunities. Universities could participate in the design of solutions and pilots at low cost for utilities, or even without any direct cost.

**Periodic asset acquisition processes.** Energy utilities often buy meters for new installations as the population grows and to replace faulty meters. The quantity of meters purchased varies from utility to utility but is generally in hundreds of thousands per year. Utilities can make good use of these regular meter acquisitions to promote smart metering projects. These conventional costs can be deducted from the costs of new smart meters. Business cases should consider these scenarios to reduce the costs of technology field trials and deployments.

Beyond acting as a driver of progress, when smart meters are purchased instead of traditional ones, they also provide an opportunity for energy companies to familiarize their customers with the new technological environment. Smart meters can act as a research platform to impart a better understanding of the needs of consumers and to guide massive rollout strategies.

It is also an opportunity for other market participants, such as local authorities, to gain a better understanding of the effect of these technologies and the ability to achieve their goals. This can guide mass deployments, define targets, and analyze governance and regulatory rules needed to ensure efficiency.

Clearly, smart meters are more than a simple driver to utilities and other market participants. We explore this in more detail below.

## Opportunities with smart metering deployments

Now that we understand the drivers, we can study some opportunities for these deployments. In this section we won't explore every opportunity but will address some of the most common ones.

Energy utilities should create and maintain full logs and analyses. Benefits realization criteria must be associated with these opportunities, with continuous updates during different stages of the project.

As these projects could affect different market participants, it is necessary to discuss some of them individually. The list of participants considered here is not exhaustive but rather the minimum necessary to call your attention to the effect of these projects on different market entities.

There may be several opportunities considered for one market participant that may also, directly or indirectly, apply to others. An example of this is, of course, the utilities themselves. We will look at the majority of opportunities in this section with a utility perspective, because energy utilities are often responsible for implementations. However, that does not necessarily mean that a utility gains more benefits than other market participants. Thus, many opportunities may equally apply to customers, local authorities, and several other market participants.

The individual opportunities to be discussed here are not necessarily applicable to your specific environment; their applicability should be carefully analyzed case by case.

**Customers.** The opportunities smart metering systems can offer to customers are numerous, although it is not always obvious what benefits can be realized. Efficient deployments can encourage customers to see this system not only as a cost but also as a service with opportunities and benefits. Improved customer relations because of better communication is the first benefit these projects can offer.

Smart displays can improve the relationship between consumers and energy suppliers. As discussed, this efficient communication channel gives customers a better understanding of their consumption and budgeting.

For example, displays can allow customers to interact with utilities, thus creating an opportunity for the utility to suggest new customized products and services and for customers to provide feedback on current ones.

Among all advantages, financial benefits appear to be the most essential for customers. Demand-side management rates can yield savings. Load-shifting services can bring energy budget reductions in different ways. For illustration, smart plugs can turn off some types of equipment on standby mode during the night and then reconnect them in the morning. Intelligent heating and water-heating storage controllers can be programmed to be activated and deactivated. Other devices and appliances can be switched on and off in a similar way. These products, together with intelligent energy-efficiency advice, would enable customers to reduce their energy usage and save money.

Customers can save money through their own actions by reducing energy consumption, and the related carbon emissions, based on cost information and forecasts provided on their display units. These devices can also provide dual-fuel information. Functions can be personalized to individual needs, and flexible payment options such as PAYG rates are another advantage of these display units. Many new services and products could be offered, such as in-home automation, incentives to customers to implement micro-generation sources based on renewable energy, and green rates.

Another aspect is the potential to end estimated energy bills. Even if allowed under some circumstances by local authorities, they are potentially a nightmare to both customers and utilities. Uncertainty as to how much to pay at the end of the billing cycle and enormous invoices caused by earlier underestimated bills are common drivers for creating debt and defaulting on payments. Smart metering systems can rectify all this and produce benefits to the consumers.

These systems assist utilities to better manage their networks and improve the quality of energy supplied. This helps avoid unplanned service interruptions, blackouts, and equipment fault problems on the grid.

Massive deployment of this technology can also help utilities to reduce their energy prices. This aspect is even more significant in competitive markets. Several factors apply in highly competitive markets, such as improvement in utility costs for energy purchased, potential reduction of operational and other costs, and reduction of peak consumption. Companies can become more socially responsible and increase their fuel-efficient products and offerings.

Customer privacy is sometimes affected by utilities. The need to gather meter readings at the end of billing cycles can be inconvenient to consumers when meters are installed inside their homes. Someone must be at home to provide readings, and when this is not possible there can be problems with estimated bills. Another potential problem is imposters with malicious intentions pretending to be utility employees.

For these reasons and others, smart metering platforms offer important opportunities to customers if they actively participate during the entire project from conception to postimplementation. Customer feedback is crucial.

**Utilities.** Different branches in a utility can make good use of many opportunities in different ways; however, all perspectives are not necessarily treated in this section. Thus, it is important to ensure that all these branches actively participate to the process of gathering requirements and opportunities. This is crucial to guarantee that all benefits are achievable and included in the project.

Due to the quantity of utility opportunities, we will only discuss the top 10 here.

***Energy-efficiency advice.*** Energy suppliers can offer intelligent advice to their consumers in terms of energy efficiency (fig. 3–10). Smart metering systems can advise customers by providing an analysis of their behavior related to energy profiles, rates, and other aspects. Other systems, such as meteorology processing, are able to provide external online information. These data can be matched to utility data and enable prompt advice to consumers. Weather conditions such as temperature are strongly related to consumption. Online advice, forecasts, and demand-side management products can make good use of this correlation.

Cross-analysis with personal customer data offers more opportunities. Consumers may want to invest in products that provide energy savings, but they need detailed information before making decisions. For example, what would be the best investment: to improve insulation levels in the home in order to increase efficiency of their heating systems or to change their incandescent light bulbs to energy-saving ones? The answer varies, depending on the type of property and consumption behavior.

In order to assist consumers, some utilities benefit from smart metering platforms by offering energy-efficiency advisory tools. These products match customer-load profile information with personal information. Private data can be obtained, for example, from a few key questions asked of consumers through channels such as websites or in-home display units, taking into account the legal requirements and the agreement of the customers. These questions are usually about the home's construction, customer behavior, and related topics. The questions should be able to be answered in only a few minutes so as not unduly delay consumers. Therefore, the questions must be structured.

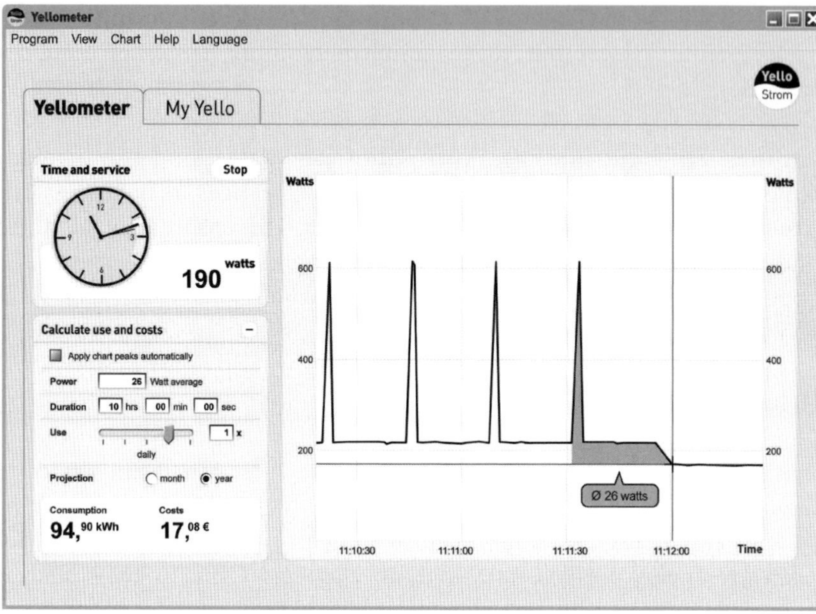


Fig. 3–10. An energy-efficiency advice platform integrated with smart metering systems. (Photo courtesy of Yello Strom.)

Relating these questions to energy-load profile data could result in customized energy-efficiency recommendations with financial implications. This report would inform customers of the more appropriate action to be taken. The recommendations should be ordered according to greater investment returns. The document should detail the investments needed (taking into account cost of capital and other financial parameters), the time required to realize savings, and a list of professionals able to deliver these services, if required. These recommendations could vary in ways such as the following:

- Replacing light bulbs
- Purchasing energy efficient appliances
- Installing additional insulation
- Rate advice

- Changing behavior (for example, running appliances at night when rates are lower)
- Demand-side management propositions
- Relocating appliances for improved efficiency (for example, in areas with low temperatures, moving equipment such as a water heater from a cold external location to inside the home)
- PAYG

Currently, there are companies that provide these services to customers; however, energy utilities seem to be much better placed for the job. Intelligent data analysis can lead to key opportunities in this vein. Utilities could also partner with energy-efficiency companies, consultants, in-home service providers, and appliance manufacturers in order to serve customers. Different kinds of businesses can be set up with these enterprises. It is a win-win—the customers would gain because the recommendations could easily benefit them financially, and the utilities could benefit from partnership agreements.

It is also important to mention that energy efficiency is more than just reducing energy demand. The tendency to return to high energy usage, in another words returning to the old habits of energy consumption, should be avoided. Consequently, actions to change habits should be implemented in a way that customers truly understand and become committed to. Otherwise, the demand could quickly revert to its original state. For example, it is common to see customers engage for an initial period when smart meters are applied, but after some time their energy-efficiency actions lapse. To make a health-related analogy, we are not talking about a diet but a nutritional reeducation. Because of the presence of temptations, it is often harder to maintain reduced weight than to lose it initially. Similarly, technology involves comfort temptations. For example, energy savings could be replaced by increased heating consumption. If customers have a habit of adjusting their thermostats to levels that save energy, reduced rates might tempt them to increase or decrease the temperature by a few degrees, which would increase energy usage. It is not a matter of teaching customers how to spend their money, but explaining to them how this action would affect the objectives of reducing consumption and consequently helping the environment and society as a whole.

This education process should involve all of society—including children. Energy-efficiency programs at schools are very helpful here. For example, there are utilities in Europe offering awards to students and teachers who reduce their energy consumption.

Opportunities seem to be innumerable in this area. Creativity and intelligence are essential to maximize benefits to utilities and other market participants.

***Reduction of operational costs.*** Normally, utilities have difficulty identifying waste. This could be due to their lack of process automation, such as metering. Smart metering implementations can help identify losses in different areas such as revenue protection and field operations.

Operational works are an example of these challenges for utilities. Besides producing technical benefits for enterprises, the reduction of related costs often provides benefits to other participants like employees and customers. For example, employees could obtain financial gains such as increased bonus related to a company's production or a reduction in price of energy.

Smart metering systems can help utilities to reduce costs in several areas, such as site visits. One site visit to read a smart meter located inside a customer's home might cost more than the device itself. With these deployments, as meters are remotely managed, utilities do not need to visit consumers to carry out the following activities:

- Meter readings for billing cycles and other needs such as change of supplier
- Simple meter tests
- Extra appointments, for example, due to meters installed inside the home and absence of the consumer when needed
- Installing power limitation devices or fuses. Local authorities sometimes require these so that consumers are not immediately cut off for nonpayment. Replacement of fuses is frequently needed when consumers exceed their limits of allowed demand. Thus, these costs could be significant as a result of many site visits.

However, these procedures could be remotely executed and their costs fully eliminated. There are further points where the need for visits and the associated time can be reduced, consequently yielding savings:

- Local work to execute remote disconnection and reconnection events
- Work on remote consumer premises on islands, in forests, and rural areas

- Local readings for rebills
- Optimization of field visit time, for example, elimination of time waiting for customers to come back home for an appointment or to authorize technicians to execute work

Smart meters are able to operate switches remotely for different needs, such as when current levels are incompatible with a predefined profile, or to execute remote disconnections. This would significantly reduce costs.

Another area of savings is the call center. Reduction of calls is expected due to new functions that smart metering platforms can offer. For example, complaints about energy quality and outage alarms can be transmitted through the metering system, eliminating the need for customers to call. Customer display units also offer new channels to schedule appointments for field services.

These platforms add opportunities for managing customer surveys, electronic bills, and delivery of customized messages. This could reduce, for example, the number of outsourced services for fieldwork and correspondence delivery. Indirect costs related to these procedures, such as invoice printing, could also be reduced.

Functions that monitor energy quality, like flicker and harmonics, can aid preventive network maintenance and management. Meter burden could also be reduced and consequently the level of technical losses and associated costs.

Prepayment service upgrade to PAYG tends to reduce costs relative to infrastructure such as reduction of pay points, eliminating the need for card or key management.

There are many other ways to reduce costs for energy utilities, which will be studied later in this chapter. They are not explored here because they relate to other benefits that need to be highlighted and studied in detail. Examples are reductions of revenue protection costs and network management.

***Reduction of peak consumption.*** This is another big area of opportunity. The daily consumption profile of energy utilities around the world is not flat. Consumers have different needs during different times of the day. The consumption pattern is usually predictable but becomes a problem when there is a huge difference compared to the average.

Peak points are common in a daily profile. The challenge is to try to reduce the differences of energy consumption between high, middle, and off-peak periods. To achieve this, load management is paramount.

Sometimes, despite the energy-efficiency actions previously described, demand does not decrease by itself. Thus, peak times can represent a real risk of energy collapse and blackouts. Unfortunately, for emergencies, it may be necessary for the distributor to promptly reduce demand. This can be part of rate arrangements or required due to an emergency via a code red or similar alarm. Smart metering systems offer different ways to do this on the utility side of the load, such as:

- Load curtailment: In this service, the distributor can send orders to groups of meters to enable disconnection and reconnection at a defined time. These orders are normally executed gradually along the network under broadcast commands. Reconnection procedures must be safely executed.

- Power limitation: Smart meters are able to offer power limitation capability. This can be part of a rate or be used during a network emergency. Once a predefined profile is established, the customer could be advised via alarms about exceeding consumption. Consumers then intervene to reduce their load. If the demand is kept at the same level, during a defined period, the customer is temporarily disconnected. Once a time window is passed, reconnection is safely enabled by the metering system.

These methods can produce efficient results for distributors and sometimes to suppliers, although they might be avoided due to the aggressive way that orders are commonly executed. Thus, it is recommended that they be used only as last resort. For more advantageous results for all the market participants, demand-side management is the best way of providing the necessary reductions. Efficiency is achievable because of the deliberate agreement of customers to participate in these programs and gain benefits. Beyond providing benefits to utilities, they reduce energy bills and have several other advantages to consumers and to the environment. This is discussed in more detail in the next section.

***Demand-side management.*** Energy demand has been steadily increasing because of higher demand for appliances for comfort, status, and other reasons. People sometimes have five televisions in their home, for example. This behavior is good for business interests but has increased energy consumption in the world to unmanageable levels. Something must be done about the quantity of energy demanded by society in order to avoid catastrophe. Beyond manufacturers taking action to reduce unnecessary consumption, such as standby and other loads, demand-side management seems to be an efficient way to deal with this challenge. It gives the ability to reduce demand at the source. There are many products that deal with demand-side management. An interesting one is load shifting.

Load shifting allows appliances and other in-home devices to be allocated special rates. Equipment can be switched on or off in defined time windows during the day. Many devices can be part of this scheme, such as electrical heating systems and boilers, and they can be switched on and off using different methods. In the past, one-way ripple controls and RF broadcast devices were used for this purpose. Nowadays, smart metering systems make this solution easier and enable more efficient and flexible management. For example, a utility can stipulate that rate signals will be sent to the meter from 1 a.m. to 7 a.m., and during this period, the energy price will be reduced for four hours. During this seven-hour window, rate orders could be sent at different times over several short periods that would add up to four hours per day. Using these signals, an in-home metering gateway would be able to command external or integrated contacts. They can switch through a boiler controller and enable it to switch on or off according to the energy product. Everything is done automatically to decrease energy prices to the customers without interfering with their comfort.

Smart plugs are an example of state-of-the-art technology (fig. 3–11). Simple versions that are available on the market are commonly run by a traditional remote control device. In this case, a simple press of a button enables smart plugs to be turned on or off locally by consumers. This is an opportunity for customers to reduce their demand manually, via their own management device. However, it seems impractical for the consumers to deal with this task every day. A more practical solution is to integrate these local smart plugs with smart metering systems.

Fig. 3–11. A smart plug. (Photo courtesy of PRI.)

In this scenario, smart plugs could work similarly, but based on rate signals. Thus, they become communicating plugs with embedded switches, and sometimes with an additional internal submeter, able to be commanded by an in-home gateway such as one designed for smart metering purposes. So, a metering gateway can command these smart plugs according to a specific rate or network condition. Additionally, they can be used to avoid standby loads for devices such as transformers for recharging mobile phones and video games. Sometimes, these are left permanently connected in the electrical sockets without any purpose. Of course, even without a transformer charge, they are consuming electricity from the grid.

For even more customer convenience, smart plugs, and consequently any connected appliances, can be remotely switched on or off via a web-based platform. Thus, a consumer would be able to turn on appliances such as heating systems before arriving at home from a computer or smart phone. This is an improvement over traditional heating controllers, which operate according to a preprogrammed schedule. This can reduce unnecessary energy use on days when we arrive home later than usual, for example. The possibility of turning devices on or off remotely can save significant amounts of energy, taking into account the fact that some devices require high levels of consumption. Safety, of course, must be a primary concern for these options.

Home automation panels can offer similar opportunities. Touch screens can give local orders to these devices. Mobile phones and similar devices can be used similarly for this purpose. The association of automatic tasks and manual operations can help customers deal with exceptions and to put in practice new ways to save energy that, if proven successful, can be then implemented as an automatic task or rule.

Because of interoperability features, appliances themselves can be directly controlled by smart metering systems and be switched in accordance with rates. They can be turned on when energy rates are cheaper.

All these methods are associated with real-time prices. In this case, options could be given to customers to accept a temporary instantaneous rate in favor of a reduced price for a defined period. Acknowledgment of this service could be made via the display interface. This rate could be enabled days or even hours before it effectively applies. These arrangements could be used for different reasons such as networks needs and better energy prices available on the spot market.

Display interfaces can also help. They could offer customers consumption forecasts, monetary dual-fuel information, synoptic analysis about rate price and consumption, and energy comparisons. They can offer consumers the possibility of locally configuring demand-side management parameters such as data security, times when devices will be turned on and off, as well as priority rules for disconnection of appliances in the event of a rate signal request.

Another opportunity is distributed energy storage. Batteries are provided to offer energy storage during off-peak periods, to be used during peak times. This method can be implemented in several ways from providing installation of new devices for DC-AC conversions in the home to using existing sources of battery storage, such as plug-in vehicles.

Renewable sources of microgeneration could also be used during peak times. Smart metering systems can optionally manage these devices or simply enable integration with energy utilities by measuring import and export energy and reactive energy and by providing interoperability with submeters. These options, beyond optimizing network consumption, help utilities to achieve energy targets of improved renewable generation sources. Other indirect action can be taken, such as installing heat pumps in order to optimize the use of thermal energy in the premises and consequently reduce energy consumption.

***Optimization of network management and quality of supply.*** Smart grids provide services such as network self-reconfiguration, automatic fault isolation, fault detection, and sensors to detect maintenance needs. Smart metering systems can also provide many interesting opportunities here such as basic network management. Both can help utilities reduce their operational costs.

Electricity meters can offer diagnosis and reporting of outage events. Intelligent algorithms can help to determine the area affected by blackouts and guide corrective action. This can reduce the time needed to solve problems in the field and also reduce the number of calls. Automatic voice messages can be promptly activated in call center systems before customers call to complain. Voice messages would inform the consumers that the utility is aware of the problem and has already begun efforts to solve it. In this case, the calls would be initially held by an automatic answer system, avoiding the need for staff intervention, except when really needed. In the gas utilities, network pressure management could also be optimized through these deployments.

Anomalies on the grid can also be promptly detected and corrected to avoid affecting premises, such as over- or under-voltage events. This process can reduce costs to energy utilities, such as for correcting equipment faults from network anomalies, as well as avoid fines charged to utilities for failing to meet regulatory targets for the quality of energy supplied.

Power factor correction can be identified via intelligent analysis of profiles from customers and meters. This could enable the installation of specific equipment along the grid to correct these levels, such as capacitors. Another potential solution is the possibility of applying special rates to educate customers about these problems. Rates related to reactive power have been implemented by different countries around the world. They enable customers to improve their individual power factor. Values below a predefined threshold can result in penalty fines and higher energy bills to consumers. These actions are essential to reduce costs related to network investments necessary to deal with these challenges.

Another point is safety. This paramount requirement can be improved with implementations of smart metering systems. For example, gas leak detections are possible. This functionality alerts customers and isolates the gas supply when these problems are detected, for example, due to people forgetting about an open valve on a cooker. This service is often available for reconnection events, but it can also be offered for normal daily operations via the installation of interoperable gas detector sensors. In case of detection, the meter valve is opened promptly, and after a predefined time, consumers can safely reconnect their supply (for a limited number of attempts). Automatic reconnection is not recommended. Thus, this procedure is only possible via local acknowledgments on the meter or on customer display units. The consumers could be warned by display messages with basic information about how to deal with this event and, if necessary, told to contact the technical call center where available specialists could advise them on how to solve the problem. For example, consumers could be asked to close any open valves in their premises, such as in their cookers, and then retry a supply reconnection. If problems persist, meters would block the reconnection function and require a visit of technical expert to the premise. It could only be reconnected after an inspection by the energy utility. Safety aspects must be taken into account during all related procedures and advice in order to ensure zero risk to customers and their properties.

For electricity grids, safe reconnections are fundamental too. Meters can receive customer acknowledgments, before being reconnected, in order to avoid risks to consumers and their premises. As for gas, customers can enable reconnection events via meter buttons or using their display interfaces.

Another point is dispatch of energy. Load simulation and emulation systems can be used to provide load analysis via energy profile data from customers. Online analysis of load profiles can identify points of problems on the grid for prompt or future correction. This could result, for example, in temporary network reconfiguration or managing transformer overloads during peak times.

***Reduction of nontechnical losses.*** Although this is extensively discussed in chapter 1, we are going to study some extra opportunities that smart metering platforms can offer to deal with this challenge. This area is usually the main payback input for deployment business cases.

Smart metering rollouts can assist utilities to identify their levels of nontechnical losses individually per substation feeder. Once a challenge to distributors, this is now possible with the installation of energy balance meters to measure substation feeders or individual power transformers. During meter installation, if all the premises supplied by transformers have smart meters installed approximately at the same time, then online nontechnical loss analysis can be done. This is crucial to ensure that there are no abnormal losses associated with the transformer when the field installation is complete. If any irregularity is identified, it is promptly fixed while the technicians are still there, avoiding the unnecessary cost and inconvenience of revisiting premises. The system is configured with the values of technical losses related to the circuit. Therefore, it automatically detects anomalies if higher values of losses are detected. These values must be carefully investigated and defined, case by case, as circuit levels can be very different. This procedure also reduces technical losses, for example, via replacement of cables and connection terminals if necessary during installation. Periodic energy balancing should therefore be carried out, for example, four times a day. These analyses generate field inspections if irregularities are detected.

Flexibility is paramount. Systems should be able to provide customized basic calculations of load profiles and meter registers. For example, a system user could request software to sum the curves of 100 customer meters and subtract this result from the curve registered on an energy balance meter connected to the transformer that feeds all those premises. This would create a loss energy profile, which could then be converted to a percentage if needed. This flexibility of adjusting the curves using settable parameters and mathematical operators is essential for revenue protection action and can be achieved thanks to smart metering platforms.

Besides detecting tampers, online analysis helps to eliminate them quickly in the field. There are costs associated with energy recovery procedures and loss prevention activities, for example, for the resources necessary to calculate penalty fines, manage appointments, and negotiate financial agreements. Thus, it is better to ensure that tampering is quickly resolved. Lower values of losses are much more easily recovered. Prompt field action also acts to dissuade consumers from attempting to tamper with meters. For example, if someone is caught in the very act of energy theft, it is unlikely neighbors would attempt it, at least for a while. Metering management centers are crucial to these analyses, action, and feedback. Implementation requires investment in equipment and specialized staff; however, efficient corrections easily pay for these costs.

It is impossible to combat a problem if you do not know how big it is and where it is. Smart algorithms can be implemented to identify the exact premises where tampering is occurring. Technicians are then guided to the correct location where the irregularities are found, avoiding lost time and costs. Of course, accuracy is essential.

Return on investment for individual action must also be evaluated and managed. Sometimes, an inspection can be more expensive than the money to be recovered. Prioritizing action is essential to obtain the best benefits from revenue protection action.

These systems also improve asset management in the field and in utility databases. Automatic algorithms, beyond analyzing load profiles to detect nontechnical losses, can also be used to detect system inconsistencies. For example, a low-voltage customer supplied by an indirect metering system (e.g., using current transformers) found with a conversion factor equal to 1 in the billing database is a case to be properly investigated in the field. It is important to mention that sometimes utilities spend time and money on high-cost nontargeted field action while problems could be better coordinated with simple database investigations and smart algorithms to guide field inspections. Integration with other databases such as weather forecasts could also be used to orient these campaigns using cross-data analysis. These platforms then become efficient tools used for strategic decisions on revenue protection programs.

Meter replacements offer ways to improve the accuracy of metrology. Smart electronic meters are able to provide improvements compared to electromechanical meters. These replacements can also help with database scrubbing and can thus serve as a reregistering activity. For illustration, it is common to find entries in databases for installed meters that have already been removed from the field. This creates additional management costs for these nonexistent assets and potential service payments to manage these "phantom" meters. Smart metering platforms also ensure that these problems are not reproduced with plug-and-play installations and authentication processes. Unbilled customers can be detected and nontechnical losses saved during rollouts.

Note that there are also some innovations in this area that could be associated with smart metering systems to reduce nontechnical losses. For example, in Brazil some utilities have tried a new power transformer with integrated measurement (fig. 3–12). Its primary side is connected to the medium-voltage network, as usual; however, its secondary side is fully measured by a fiscal meter (energy balance). Additional specific meters also measure the low-voltage distribution cables used to supply the individual customer homes. Thus, all measurements occur at the transformer level, which limits the access of people who may want to tamper. Customers can check their consumption information through a display installed inside the home. The specific compartments where the meters are located are made of resistant material and monitored online using special-purpose sensors. Thus, any intervention could be promptly communicated to the middleware.

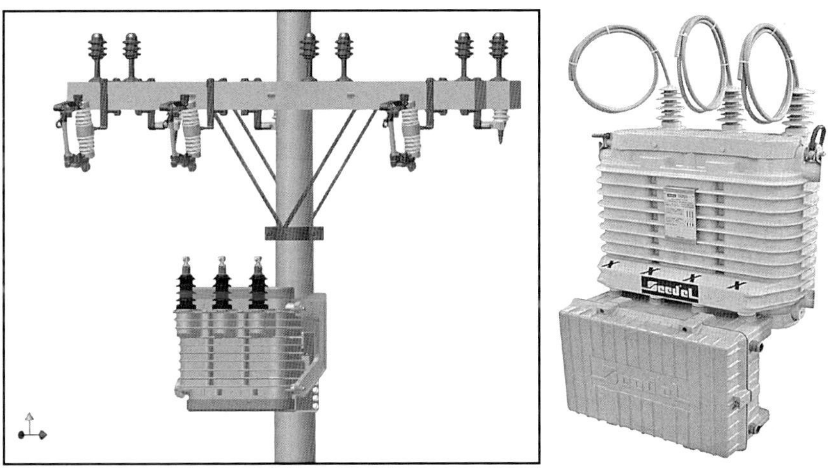

Fig. 3–12. A power transformer with integrated measurement. (Photo courtesy of Seedel.)

***Financial management.*** Financial management can produce savings for energy utilities by optimizing cash flows. Although the majority of customers pay their energy bills on time, late payments are common. This creates monetary problems for utilities and extra costs for rebilling, field visits for disconnection and reconnection, and legal procedures that eventually become necessary to recover these amounts. Sometimes, the costs outweigh the value to be recovered. When allowed to accumulate, they create a debt management need that can involve appointments, negotiations, and other actions to the point that they are unlikely to be paid by customers, even if divided into regular payments.

Among the other solutions that can be offered by smart metering systems, PAYG seems to be the best one to help customers to be better prepared to face these problems. PAYG also improves the existing prepayment system, for example, by avoiding the costs of reconciliation in competitive markets. These systems offer flexible channels and payment methods to prevent accumulated debts by customers. If necessary and unavoidable, meters can also be disconnected remotely, which eliminates the cost of site visits.

Another point is ending estimated bills. As already mentioned, apart from creating cash flow problems, they create debt due to unexpected, expensive bills.

Bad decision making generally creates financial problems, and good information is needed in order to make wise financial decisions. For example, utilities can be faced with questions that in theory are easily answered, but harder in practice due to absence of decision tools and good information. The following are examples of these questions:

- For which customers would it be more beneficial to invest in revenue protection: residential or industrial?
- Where are the biggest levels of nontechnical losses? What regions would provide the most benefit?
- Is it better to invest in revenue protection, differentiation services, or PAYG?
- What are the services that customers would pay for?
- What are the real levels of nontechnical losses in relation to total losses?
- What are the values for energy balancing that bring prompt action in the field?

Data-mining systems can work through smart metering data warehouses to provide information to be used for analysis of investments and help with decisions. Queries could also be used, in association with field research, to target products according to customer needs and added-value perception.

Smart metering platforms are important tools that can improve financial management in energy utilities.

***Customer retention and competitive differentiation.*** Retaining a customer is much more economical than attracting a new one. Losing customers, in addition to effects on market share and sales, leads to costs to acquire new customers to cover the gap. Thus, improving loyalty levels is key for utilities operating in a competitive market.

Loyalty is often connected to customer perception of value for energy services. Energy price is important, but other parameters exert a greater influence on customers, like good service, innovative products, and trust. Smart metering platforms offer opportunities to customers to interact with their energy utilities in a different way. They enable energy to be considered not only as a simple commodity but also as an added-value service.

Innovative products, technology, and efficient strategies, together with other benefits that smart metering platforms can offer, are key to this evolution. Energy-efficiency action could also contribute to this process, if well implemented. After all, who would not like to have their energy bill reduced? If someone could provide this benefit for you, and if you were sure this benefit would be sustainable, would you change suppliers? What about the goodwill these successful actions create? Word-of-mouth marketing is still the most efficient and economical way of attracting new customers. In summary, customers perceive these products as added value and want to continue with their energy supplier. Before investing in creating new consumers, utilities must ensure that their customers are satisfied. Beyond differentiation, even though smart meters enable customers to change suppliers more easily, smart metering can also promote customer loyalty!

Providing new services is important to utilities to survive in aggressive markets and improve customer satisfaction. Despite customer retention strategies, increasing marketing share is often a major target, too. Smart metering platforms can provide the necessary products to enable utilities to achieve this target. Differentiation services can be offered via the large number of features this system enables but also through integration of this platform with other products. The following are examples of opportunities:

- In-home automation like lighting controls and smart locks
- Remote command of smart plugs and appliances via the web, based on both scheduled and on-demand orders
- Existing in-home display interfaces
- Home safety services like flood, smoke, and gas leak detection
- Integration with cooling systems and heating and hot water controllers

- Weather forecast displays and analysis
- Internet, video streaming, and VoIP services
- Creation of energy service companies (ESCOs)
- New income streams from partnerships
- Health care devices like HAN-based diabetic testing, panic buttons, automatic blood pressure monitors, and electrocardiogram routers
- Home security

The above opportunities can be customized according to the utility environment. While in some countries peak consumption is mainly related to the use of heating devices, in others these levels could be related to air conditioning or even to both systems, depending on the season. Despite these customizations, these products enable utilities to change customer perception of their capabilities. Energy suppliers now sell energy efficiency and comfort to their consumers. Improvements in reputation and company image are an indirect benefit of this perception.

Partnership resources should be used for services not related to core business of utilities in order to ensure optimum performance of products offered. Use of specialists can be more appropriate because of their expertise and experience.

Customer relationships have entered a new era with the use of innovative display channels. Marketing can be improved through these displays. Utilities can now profit from these devices to deliver messages to their consumers for different purposes, such as offering new products. Subject to legal requirements, these services could also be offered to partners such as insurance and energy-efficiency companies. Advertisements and other marketing tools could also be offered in a similar way.

Another point of improvement is optimization of customer relationship management (CRM) platforms. This new display channel can enable reduction of costs and opportunities for different related functionalities such as:

- Multilanguage interfaces
- More frequent customer surveys
- Local energy credit acquisition for PAYG services
- Advisory services for customers

- Rate advisors
- Interaction with consumers with special needs
- Electronic billing
- Acknowledgment of message reception
- Appointment tools
- Correspondence delivery
- Submetering monitoring
- Integrating other meters such as water and gas

Load-profile analysis can offer opportunities for differentiation. This analysis can be used to help energy companies to understand customer behavior and needs, for example, when they use microgeneration, emergency credit, and how often they purchase credit. This feedback can be used to create new products and confirm existing ones.

These analyses also enable utilities to practice social responsibility. Consumers who have not paid for electricity for some time could be identified and asked if they need social assistance. They could be directed to contact government agencies or charities to help them deal with this problem. Sometimes, financial problems make people feel depressed to the point that they isolate themselves and refuse assistance. They could be cut off during a cold spell and may not have enough resources to cope with the situation or look for assistance. These analyses can help utilities to identify these cases and help consumers in trouble.

Customized interaction can be provided for special needs customers. Customer display units could provide voice messages and special key signs to assist blind people. Several other services can be offered as well.

Finally, these services do not apply only to residential customers. Major, medium, and small businesses can also obtain benefits from implementations. Other products can be related to their specific needs like management of capacitor banks for improving power factor, industry production planning tools, and real-time demand management.

***Reduction of energy costs.*** Sometimes, generation companies are integrated with utility delivery companies, and other times not, according to market rules. However, there is an international trend to separate energy companies into fully independent enterprises with different core businesses such as distribution, trade, generation, and supply. More and more utilities in the world adopt this regime to enable competition in these market areas, among other reasons.

In this scenario, suppliers purchase the energy they offer to their customers from generation companies. The price of energy purchased varies according to the time it is needed, the source of generation used, and other aspects. In order to negotiate these contracts, utilities need to estimate their energy demands. Simulation and forecast tools are used for this purpose. Detailed analysis allows these companies to understand market change in consumption. Nowadays, these forecasts are based on total measurements, previous negotiations, assumptions, and other parameters. Sometimes, because of a lack of available information, these negotiations are inefficient and create prejudice of energy suppliers. For example, too much energy can be purchased. The opposite can also happen. The narrower the margin between energy demand and supply, the better its cost and financial benefits. Thus, taking into account that they are generally negotiated as medium-term packages, utilities can pay an expensive price for mistakes. In summary, this can oblige utilities to purchase expensive extra energy on spot markets or to deal with enormous margins of idle energy.

Smart metering systems provide data to optimize energy simulations with customer profiles and associated products that can be used to avoid mistakes in demand forecast. They can offer energy-efficiency products, such as demand-side management rates, which can enable these companies to postpone purchasing extra energy. Thus, market demand can be adapted to energy acquisition strategies and current contracts, rather than the opposite!

During peak times, in general, more expensive generation sources are used. Additionally, the extra production plants, necessary to supply the needs during this period, usually use sources that produce more pollution such as carbon emissions. As demand is higher during these periods, this price can be significant in terms of the total energy package offered to consumers and harm to the environment. Thus, reduction of peak consumption potentially reduces energy costs for utilities, usually for their consumers, and potentially for the environment. Efficient management enables, at least, prioritizing for more economic benefit and less pollution.

Other issues must also be addressed. Sometimes, the spot energy market is only used by energy suppliers in case of an emergency, for example, where there are inefficient contract arrangements with main generation companies. However, with online rates and demand-side management, smart metering systems can present opportunities. Idle energy is sometimes offered at low prices in this market. Depending on laws and market rules, this energy can be purchased by energy suppliers to offer temporary instantaneous rates to their customers. This aggressive strategy can bring several benefits to different market participants and enter a new era for competitive markets.

Finally, the benefits that smart metering platforms can provide to utilities and ultimately to their consumers and other entities of the markets are innumerable. These opportunities should be evaluated along with the financial analysis in order to guide the implementation of this technology. Of course, energy company strategies must also be respected and taken into account.

***Efficient maintenance and installations.*** Installation and maintenance must be efficiently executed in the field. Plug-and-play installation is a useful method of reducing time such that even complex systems do not necessarily require a significant amount of time to implement. Intelligent systems must enable automatic discovery processes for installation devices and must allow for simple and secure connection procedures to associate devices in the smart metering HAN. Self-diagnostic tests and configurations must be automatic and quickly executed. The use of infrared scanners and other technologies potentially also reduces the length of time spent during field visits. These procedures can link to GPS devices in order to improve asset management. In summary, in order for to these systems to be truly smart, the need for device field intervention and related time spent should be lower than for traditional metering platforms.

Predictive maintenance offers opportunities in addition to remote maintenance procedures. For example, power transformer faults are often due to overloads during peak times or problems of load balancing across different electric phases. In fact, even if these problems are seen as simple, the majority of utilities are unable to identify them due to lack of information and automation. Smart metering systems, for example, enable online fiscal meters to be connected to power transformers for measurement. These measurements can be as simple as active and reactive energy only or provide complex quality analysis involving harmonics. These meters can be temporarily or permanently installed on the grid, according to utility strategy and business cases. In both cases, they contribute to promptly identifying and correcting grid irregularities. Energy consumption and other measurements can be individually recorded per phase. Abnormal energy profiles can produce grid reconfigurations and demand-side management action. The sustainability of these achievements can later be monitored by these systems. The predictive maintenance that these new platforms potentially offer can enable reduction of unnecessary grid investment.

Alarms and logs can also be used to provide field diagnostics of potential faults in devices. If systemic faults are found, associated with specific production lots or models of meter or auxiliary measurement equipment (e.g., instrument transformers), recall procedures can quickly be put in place to correct these anomalies in the field. Investigations can also be carried out in order to analyze collateral effects these problems have created and to avoid them being reproduced. This continuous learning exercise is crucial to operational cost reduction.

Meter age can also be monitored and tied to program inspections and replacements when necessary. Remote upgrades can also be provided. Because of this, utilities are able to correct most bugs and anomalies remotely, without any need for field action. This functionality also enables devices to remain in the field for longer periods due to future-proofing of integration and customization to new market requirements. Smart metering devices themselves can automatically execute self-correction and rollback, for example, related to firmware problems. This function is crucial to avoid severe problems that commonly require field interventions such as system crashes. Efficient watchdog timers can be implemented here.

The reduction of time faults reduces related financial losses. These faults sometimes remain in the field for years without correction. At other times, they are detected but forgotten by utilities. The maintenance workflows of these systems ensure that all the detected anomalies are promptly treated in the field. Scheduling tools can encourage customers to keep appointments.

Another feature is fieldwork that can be done without AC voltage. For example, isolated ports, such as universal serial buses (USBs), can be used for upgrades in the field, such as to memory and communication modems. Special isolated compartments can be used for local replacements. For electronic gas meters, the life of a battery is often an issue handled during technical architecture developments. These meters may need to be replaced early due to low battery events, which may be caused by the consumption requirement of these smart meters to communicate frequently with middleware. It is thus common for smart gas meter business models to be found unprofitable. The use of specific isolation compartments can enable these batteries to be safely replaced in the field. Thus, because the work is not being carried out under voltage, less expertise is needed by the professional employed to execute these tasks. Consequently, intervention costs are reduced and the business cases show more benefits. For example, meter readers could be retrained to do this work and other tasks such as mandatory visual safety inspections in the field.

# Innovative Rates and Payment Systems

Beyond opportunities and drivers with international deployments of smart metering systems, innovative rates are one of the main reasons for rollouts. Of course, their implementation depends on legal requirements. In some regulated markets, local authorities can be responsible for establishing rate prices and/or structures. Even in these cases, technologies offer, at least, tools that utilities can use to offer innovative products to consumers. As the global trend goes toward fully competitive markets for energy, these platforms could create many opportunities to customers around the world. In existing competitive markets and others in the process of opening up competition, these systems offer immediate benefits to all the market participants.

Regulated markets commonly have simple rate structures for credit and prepayment customers. A single energy register is built into meters to show accumulated consumption. The related readings are linked to a simple rate and to other financial aspects like local taxes and fixed charges. This basic scenario was imposed in the past due to the limits of the first electromechanical meters. Nowadays, innovative rates can be offered to customers, providing the flexibility this new technological era requires.

In the electricity domain, these opportunities are promptly realized by utilities and put into practice once smart meters are deployed. In the gas sector, because of different factors such as time-of-use rates, some utilities do not apply innovative rates. Energy suppliers can make good use of the opportunities for modern rate profiles for gas. The cost of gas is also affected by consumption peaks, price variation according to time, and other characteristics that could be improved with these new measurement systems. We should not forget the indirect costs involved. Electric pumps are often necessary to enable gas distribution. Consumption costs vary according to the time of day and thus can be optimized. Energy companies should produce financial analyses and benchmarking exercises to put this into practice or, at least, to hold discussions with authorities to allow it in regulated markets.

New rate profiles are crucial to energy suppliers to survive in competitive markets. Several structures could be supported by smart metering platforms. The following sections cover some of them.

## Pay as you go (PAYG)

The PAYG structure is the natural evolution of prepayment systems and energy package rates. It can group the best functionalities of existing products and create even more opportunities due to its association with other products and benefits provided by smart metering systems. These platforms have the flexibility to change the payment mode remotely on meters. This means that utilities can use the same device for both credit and debit and switch between the two remotely. Until recently, when using traditional electronic meters, utilities were required to replace the meter in the field to provide this flexibility. The majority of traditional models of prepayment devices cannot be used for postpayment applications. Even when they are, there are usually intrinsic limitations. Thus, this flexibility reduces the cost of these implementations.

Additionally, although some data processing is still done in meters or in-home metering gateways, they are not fully centralized in these devices anymore. Data processing becomes more centralized in the middleware. This is possible because meters are now part of an online environment. Actions such as disconnection due to insufficient energy credits can be ordered remotely instead of locally, and the same is true for other innovative and/or process-intensive actions. If real-time tasks are necessary, the middleware can count on the in-home metering gateway, its "management arm" in the field.

Another point is that PAYG services are not necessarily prepayment services anymore. They are much more flexible. For example, energy and related taxes can be paid in advance while other charges, such as fixed ones, could be paid later. These charges could be paid at the next credit acquisition procedure, taking into account the time passed between the previous purchase and the new one.

Thus, this method destroys some concepts associated with prepayment systems, for example, that these payment methods could offer disadvantages to fuel-poor customers because they must pay their energy bills in advance. These devices are now able to communicate remotely with middleware systems at regular intervals and provides for the possibility of integrating these platforms with other systems and different services like cable TV. Initially, you could imagine these companies would never have a relationship with energy utilities, although now, subject to contractual arrangements, smart metering systems could be used to collect charges for these services. These charges could be collected in advance or after use, in order to help customers with their budgets. Among other benefits, this could avoid the creation of debt to these service providers or the costs of implementing complex customized systems. Depending on the agreements, energy companies could receive fees per transaction or other gains. These companies could legally trade this service with the partial use of the cable media band to enable communication for smart metering devices. A virtual TV channel could also be assigned to energy suppliers to be used as customer displays and provide consumers with information about their energy consumption and several other services via a simple touch on their TV remote control. Messages and alerts, when required, could also be displayed on the television and allow local action or acknowledgment. This was just an example of many other win-win partnerships that could be established via PAYG arrangements. Note that they must respect financial rules and other legal requirements like the ones related to taxation of services and proper invoice arrangements among partners.

PAYG systems offer even more flexibilities such as when combined with advanced rate structures and load management. Time-of-use arrangements can be associated with this payment method. Thus, credit would be decremented under different rates depending on the energy price active at the period the consumption occurs. In terms of load management, prepayment systems usually offer this functionality via a power limitation function, but even if it is able to enable some opportunities, it can be criticized for the inconvenient way it is provided to consumers. For example, it can interrupt the supply of energy to customers for temporary periods when power limits are exceeded. With smart metering systems, demand-side management can be provided with PAYG and offer even more opportunities than when in credit mode. Subject to legal requirements, these could be reverted to energy credit bonuses to be used by consumers. Another option is the accumulation of points with marketing promotions. Accumulated points could be used with partnership networks with supermarkets, appliance stores, airlines, hotels, and other businesses. Similar functions could be enabled with micro-generation incentives. The smart meter could measure the energy exported to the electricity grid, while the total energy created by the customer would be measured via submeters. These devices could be connected to the metering gateway through the HAN.

Dual-fuel arrangements are another area of opportunity. Generally, debit customers, when supplied by both gas and electricity, have two different meters and related registers. This is hard to manage for both customers and utilities. For safety reasons, it is unlikely that both meters will be physically integrated, but they can be logically integrated to facilitate a consumer's needs. For example, a common energy credit could be provided integrating both electricity and gas. This energy credit could be distributed online between these single fuels or separately managed if desired by consumers. In the second case, even if separately managed, the customer could be provided with a virtual combined view. When the management of the energy credit is fully integrated, customer displays would continue to show information related to the individual fuels so that customers could react to fuel-specific energy efficiency targets. Automatic advice could follow in the same way.

Payment channels are also even more flexible. Display interfaces could be used to enable energy credit payment and also payments for nonenergy services on behalf of potential partners like insurance companies, water utilities, and other enterprises.

Smart meter arrangements with PAYG enable even more flexibility in terms of rate profiles, associated devices, and many other points. For example, subject to local metrological rules, smart plugs with built-in measurement and switch capabilities could be additional applications for smart meters. They would enable similar rate discounts per electrical socket, depending on the load connected to them and the time they are used. Of course, due to all the flexibility these systems can offer, the number of rate-related opportunities is innumerable.

In summary, clearly prepayment and PAYG are two very different services. PAYG incorporates prepayment and is only enabled via smart metering platforms. It combines the best functions of debit and credit modes with different energy products to create new opportunities for customers and utilities.

## Advanced rates

Different rate structures can be created in smart metering systems, bringing together customer needs and energy prices. Both parameters can be combined by energy utilities with energy costs and fuel availability for different periods of the day. These rates can also be combined with bonuses and discounts. They are not restricted to a specific arrangement, so they can apply to many rate profiles if desirable and legally authorized. In general, they can be managed by billing engines and apply to any one of the existing structures we discuss next. In a similar way, multifuel arrangements can also apply to the large majority of rate structures. Let's now discuss some of them.

**Time-of-use rates.** This structure is commonly used around the world to bill large, medium, and small businesses. The trends indicate that all customers, including residential, will be charged using these structures sooner or later. Electronic meters and registers have made this possible for quite some time, and in some countries they have been adopted for residential customers. Different prices are associated with different times in this rate structure (fig. 3–13). For illustration, three peak times could be established during the day. Nonworking days and holidays could be treated as off-peak periods. The three peak periods could have single or different prices and a different price for off-peak. In this case, for metrological purposes, one individual register must be dedicated per rate. To offer even more flexibility, the holidays and weekends could be rated differently from the off-peak periods during the week. Middle-peak rates could also apply in a similar way to provide even more flexibility.

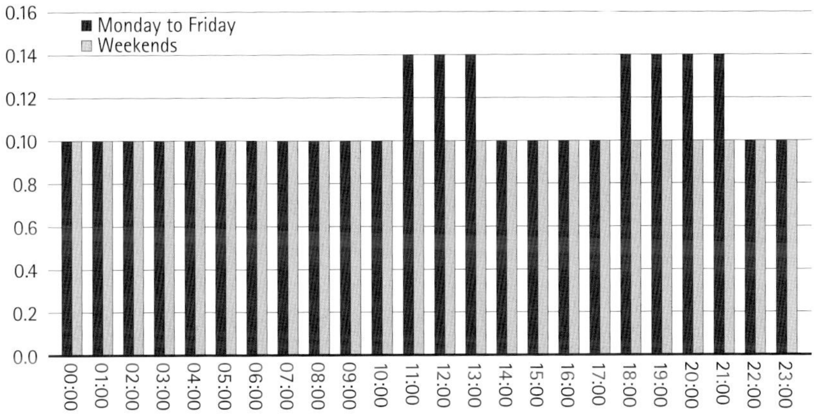


Fig. 3–13. An example of simple time-of-use arrangement

Clock settings are vital accuracy requirements for hourly seasonal rates because charge rates are associated with time. Adjusting clocks for these meters has been a nightmare for energy utilities that employ the traditional electronic meter with a built-in clock. In this case, technicians must regularly set these parameters in the field and for special events such as daylight saving time changes. These interventions are very expensive. Some other utilities have opted for using meters without internal clocks. Ripple control or RF broadcast system are normally used to send commands during predefined intervals throughout the day to switch registers and the associated rates in meters. Even if this method does not require interventions in the field for clock-setting purposes, it still creates much inconvenience to utilities such as the costs to maintain these services in operation and technology limitations on the number of commands utilities can send to the meters and the respective rate arrangements that can be used. With these technologies, clock synchronization can now be remotely set with smart metering systems. Thus, smart metering platforms enable these rate arrangements more naturally and easily for all the involved parties. It is possible to interact with customers because of optimizations through two-way communication, remote upgrades, and innovative methods.

Fixed charges, like maximum demand customer allowance, can be managed by these profiles. Different taxes can apply to these charges and to energy consumption and be related to municipal, state, and other collection entities. Because of the level of complexity, these rates are often managed by billing systems in the corporate IT environment, and traditional meters do not usually provide customers with the details they need on these rates. Local management carried out by meters generally requires frequent reconciliation with these systems, which is difficult to ensure without online interactions. This can be more easily carried out when using smart metering platforms.

These structures can be set for seasonal pricing, as rate levels are directly tied to generation costs, which can be affected by seasonal conditions. The different rain levels that occur in dry and humid periods, for example, directly affect hydroelectric power stations. These time periods then relate to different rates.

As mentioned, sometimes the demand achieves critical levels in networks. In order to help utilities face these challenges, time-of-use rates can be associated with these emergency periods. Price-signal events are associated with different periods of the year and different times of the day. These seasonal periods can be assigned colors or other graphical formats and be posted online using the display channels offered by smart metering systems such as wall panels, e-mail, IT messenger tools, web widgets, multimedia message service (MMS), and short message service (SMS). Consumers usually receive rate plan documentation that explains these periods in detail. Consumers are informed, via rate signals, days or hours prior to the change of rate. Metering gateways can efficiently manage devices such as smart plugs and heating systems according to the current rate.

Say a customer has a time-of-use rate, and every day the periods from 10 a.m. to 12 a.m. and from 5 p.m. to 8 p.m. are considered to be peak, and the other times off-peak. In terms of seasons, four colored annual periods could be defined:

- From July to August would be the **yellow** period.
    - From December to March:
        - The first half of the month would be the **orange** period.
        - Of the last half, five days could be chosen to be part of the **red** period, depending on the utility needs.
    - During the other months, the period would be **green**.

Respectively, the green period would be cheaper than the yellow one. The orange would be more expensive than the previous one, and the red the highest price of all. The red rate signals would be sent according to critical consumption periods on the network. This is necessary to balance the demand with available energy generation capabilities. The yellow and orange periods could be for generation cycles or seasonal stations. Customers would receive alerts, eight hours before events, on their display unit, web page, and optionally via SMS, MMS, or e-mail. Different colors would be displayed on these interfaces according to the period. The smart metering gateway could also be programmed to manage customers' load locally via the external HAN controllers.

Maximum demand allowances can be associated with these rates; thus different values could be allowed depending on the period of the year. If not respected, they can result in fines paid by consumers that can vary according to the period such as peak, off-peak, specific days, or annual seasons.

Several other rate structures are possible, depending on the market models and creativity of energy utilities. Efficient communication campaigns, registers, channels, and interfaces should be used to avoid misunderstandings and allow customers and utilities to obtain the best benefits.

**Flash rates.** Flash rates are temporarily applied for customers depending on specific conditions and customer contracts. These conditions can vary according to idle or on-sale energy on the spot market, network conditions, or contracted packages. They are dependent on local laws. For example, consumers could receive a message related to a sale to take place later that night. The discounted rate could be for a six-hour duration during the off-peak period. Different online display channels could be used for this purpose. This option could be included in the contracts of customers and their energy suppliers. It could also be used for some specific periods of the year.

Large companies are already able to participate directly in energy auction markets in some countries for medium- and long-term arrangements. Based on this, it is reasonable to imagine that, with smart metering systems, online short-term auction markets could be directly available to medium and small business and even to residential consumers in the future. These could be associated with annual, monthly, daily, or even hourly-based contracts. For example, a consumer could purchase 1 MWh to be used during the following week.

Smart metering platforms play a core role in these innovative offers. Online communication capability and display interfaces are essential for these services to be offered to customers and to improve differentiation in competitive markets.

**Reactive energy based rates.** Reactive energy is sometimes a serious problem for network management. It requires expensive correction across the grid, such as installation of capacitor banks. For this reason, many utilities around the world offer rates that associate energy bills for large business with the power factor conditions of their premises. For example, a power factor of 0.92 can be established as the limit of capacitive and inductive energy in the grid. When the fixed limits are exceeded, the customers must pay a penalty. These rates are currently being extended to medium and small business and sometimes even to residential customers as reactive equipment in these premises becomes more common.

It is common for these consumers to install capacitor banks to enable power factor correction. This equipment can have different values of capacitance depending on load conditions during a period. They can be managed by smart metering systems according to online measurements and contracts.

**Green rates.** Green rates are becoming commonly available for consumers. They are sometimes meant for external renewable generation sources or for local microgeneration. Thus, a renewable source can offer energy to the market; utilities purchase these packages and make them available to consumers.

This energy is generally offered at higher prices than the traditional ones but with a positive effect on the environment. Renewable generation sources are also unlikely to offer an efficient financial payback to consumers. It is more a social and environmental responsibility than an improvement of budgetary commitment. Due to these financial aspects, utilities and authorities can offer different ways of encouraging and supporting these actions such as tax incentives, mortgage campaigns, and renewable credits. Microgeneration in customers' premises often involves credits as well.

These offers are encouraged by authorities to meet carbon emission targets and are an increasing trend for this century. Some further requirements in this direction seem to be imminent.

**Grouped rates.** As mentioned, some customers have special needs. One of these needs, which is sometimes hard to achieve because of legal requirements or technology availability, relates to grouped premises and their rates and billing. Examples of people who need these arrangements are:

- Consumers with more than one house
- Parents can want to jointly pay their children's energy bills with their own
- Children who wish to pay the energy bills of their parents with their own
- Landlords who would like to have a total view of all their properties
- Retail networks that generally would like a single energy bill for all their locations
- Prepayment customers with more than one property who would like to manage a single energy bill
- Industrial customers who could save money if contracted for a grouped maximum demand allowance for all their sites

Smart metering systems can provide integrated multisite arrangements through online communication methods and advanced IT functions in the middleware layer. In addition to the single views above, consumption and load profiles could also be integrated at the middleware level. Metrologically, the registers would continue to be individually measured and managed, but a total virtual view potentially adds several opportunities for customers and utilities. New products can be implemented here. As a result, the IT tools creating these views and rules should be sufficiently future-proofed to be flexible enough for evolvement of customer needs. Dynamic mathematical and logical operators are examples of indispensable functions in these systems.

# Main Challenges

As with any other kind of automation and innovative deployment, challenges must be faced and met. Risks should also be taken into account, updated, mitigated, and managed using analysis tools like risk and investment logs.

These challenges can differ according to the environment. Some of them have already been discussed in this book at a high level and will now be covered in greater detail, together with some new ones. The aim is not to exhaust the problem but to summarize some of the obstacles that can be encountered and identify how to deal with them in order to obtain the best benefits from these projects.

## Customer engagement

Privacy of the customer is paramount, independent of what kind of technology is being implemented. Press organizations in some countries publish articles about smart metering systems and the concern related to privacy of consumers. I have been faced with this question while participating in international conferences and seminars as a chairperson or panelist or when presenting cases. This demonstrates that it can be a challenge to be faced by energy utilities during deployments. Consumers often do not appreciate having their lives monitored. They can imagine these systems as a kind of "reality show."

All market participants directly or indirectly responsible for these rollouts must respect the views and desires of consumers. Laws, directives, and other rules must be respected. This process should not be viewed as mandatory. Instead, customers should realize the benefits they can obtain from these deployments and voluntarily participate.

Initially, customers will be faced with real needs and opportunities. An efficient communication process is essential to understanding their needs. This is the time to provide a specialized view on the problems to be faced and help customers understand how to overcome them. They also need to understand that their goals depend on behavior and cooperation, and that together with their energy utility, their goals are achievable. Similar cases should be presented in order to help them realize the benefits they could achieve. Education campaigns, brainstorming, personal contact, orientation, and patience are crucial to understanding needs, the solutions proposed, and the benefits they can achieve. Customer feedback must be heard during this process, and customers must continually participate at all the stages of this deployment from the very start.

The costs related to these innovative products will be directly or indirectly borne by the customers, so it is important that they provide added value. Benefits and opportunities must be properly demonstrated by the utilities, instead of just referenced. Customization can help in this process.

The customers are now free to decide on the product proposition. Even where a government mandate exists, the customer can decide not to cooperate. This can make things difficult for energy utilities and must be avoided if possible. Thus, if all the related work was properly carried out by the utility, the customer would probably realize the associated benefits and decide in favor of installing this new technology. However, even in this case, this is only the beginning, as customers will also need continual support to achieve their desired result.

Utilities should ensure services, display interfaces, and communication channels maintain customer privacy. While these methods can be used for marketing purposes and to send messages to consumers, customer privacy must be respected. Communication should not be used for undesirable e-mails, spam, or telemarketing calls. This would potentially irritate consumers instead of attracting them and ensuring their loyalty.

Some research demonstrates that a large number of customers still do not realize the opportunities to be gained by smart metering implementations. This demonstrates that this challenge must be seriously faced by energy utilities and other market participants involved with these deployments.

Mistakes like bugs and faults, for example, can halt or delay projects and affect the reputation and image of the utilities. Proper tests and internal pilots must be in place prior to any customer installations. Customer satisfaction must be at least reached, but it would be better still if it was exceeded.

Customers could have additional concerns, even after agreeing on the project to be deployed, such as:

- Is the communication medium safe? When RF technologies are used, special attention should be taken with tests to make sure that the customers suffer no ill effect. Some customers can have concerns about this, and these concerns must be properly managed.

- Will I be able to use this system? Technophobia is another common problem. Some customers do not enjoy technology and do not want to deal with it. User-friendly utility interfaces can help consumers face this challenge.

- Resistance to innovations. Science and psychology explain our resistance to everything that is not inside our comfort zone. When the challenges are managed and success is achieved, then technology enters our comfort zone and becomes part of our life. Otherwise, it can create a barrier that will be even more difficult to cross in the future.
- Who pays for it? This is another aspect to face. Customers sometimes see these projects as ways for utilities to increase their energy bills. Good communication and transparency are needed.
- As with many products, smart metering systems are normally implemented in layers. For example, the basic product could be offered to consumers at no cost. Advanced packages would require a charge to be paid. But what would customers pay for? Answering this question is essential to build these projects and related propositions. Proper research must be made to provide answers for each piece of environment.
- Should I allow utilities to use my private devices? When an energy utility uses customers' broadband Internet modems to piggyback communications for smart metering devices, it should take even more care. For example, customers could blame energy companies if their modems are faulty, even if due to obsolescence of the equipment. Efficient contractual clauses and communication process should be in place.

Finally, if concern and other challenges happen, utilities must be able to face them.

## Production chain and future-proofing capability

Producing millions of devices is clearly not the same as producing thousands. These projects can cause manufacturers to restructure their production chain to meet utility requirements and timescales. Utilities must ensure that solution providers respect proper quality and social responsibility procedures and policies. Safety is paramount and quality of products is usually associated with it. Utilities must equally avoid any kind of inappropriate employment from their contractors and partners such as below minimum wage standards, and so forth. Audits must take place to ensure that all the requirements are met and policies fully respected by all parties.

Apart from quality, other problems can also affect products installed in the field over time. Non-future-proofed specifications are the main cause of these premature replacements. This can generate extra costs for the utilities and possibly destroy business cases. Good design, interoperability, and upgrade capability are essential to ensure an efficient deployment.

In-home metering gateways must be able to download and run new applications and interact with other systems to ensure the future-proofing of these implementations. Middleware should follow similarly. Efficient specification is a critical factor for these projects.

An end-to-end analysis from the project beginning up to its full life span needs to be carefully carried out. The success of projects also depends on this.

## Business model

There are several reasons why projects do not succeed. One of them involves problems with the development of business models.

Sometimes opportunities and benefits are assumed without having been based on proper customized studies. It is unfortunately common that many projects do not recover investment in a reasonable period. This can happen, for example, because some of their benefits were not fully realized. Often the blame falls on excessive optimism and bad benchmarking exercises. Assumptions taken from other projects cannot apply from one to another, even if they are similar. However, it does not mean that we have to be pessimistic. Excessive pessimism can prevent good projects from getting started. An equilibrium point between both seems the best approach. The message is, excesses must be avoided.

False promises are another common problem. Never promise anything you cannot deliver, as this can create several problems for the utility in terms of project reconciliation and its image. Expectations must be properly set.

Note also that sometimes it is better to stop a project while there is still time rather than continuing and generating future problems for companies. Correctly halting or cancelling a project is not a cowardly action but a responsible one, if carried out for the right reasons. For example, the smart metering project many not be financially beneficial if deployed as a massive rollout. Sometimes targeted deployments are more advantageous. Another point is that one project should not be analyzed in isolation. For example, a massive smart metering deployment on its own may not be a justifiable investment, but could be if associated with smart grid or billing replacement projects.

Further advice: Do not necessarily believe all you hear, for example, when someone would like to sell you a solution. Unfortunately, this often happens during benchmarking exercises. Some utilities would prefer not to demonstrate a failure in their deployments. Reports of benchmarking exercises require attentive study and reasonable analysis. The same applies to in-person demonstrations. Everything must be properly investigated and tested. Other views can also help, such as those of customers, regulators, other solution providers, and other participants. Energy utilities must execute their own analysis to make sure that their deliverables are effectively achievable. They must do this with transparent and realistic analysis.

## Legal requirements

Not all products a utility would like to implement can be effectively deployed. This is a reality to be faced. Legal barriers block several services. It does not mean that you should give up, but that proper studies and cooperative action should be carried out before decisions are made.

Creating and participating with working groups is a common way to enable cooperation. Active participation of legal specialists from the project conception is fundamental to identify risks and issues that can affect the success of a project. Resistance should be dispelled with facts. Participation in tests and pilots generally assist the project team to gain a better understanding of the projects and the environment. Remember that sometimes people resist the unknown. They must be provided with good levels of information in order to make them comfortable with the projects and regulations that have to be established.

Authorities and other market entities need to organize transparent discussions around the market model they must adopt. The chosen model and smart metering platform must enable benefits and opportunities for all the market participants. Thus, with this collaboration, the market can deal effectively with these challenges.

Asset ownership is sometimes another challenge. Meters can be owned by the customers, by the energy supplier, or by the distributor, and this can vary according to country. Each case requires a different approach and can entail a number of specific challenges. For example, if the suppliers own the meters and they are responsible for smart metering systems, it is unlikely they will use the cable infrastructure with a PLC-based technology in a fully competitive market (unless required by authorities). This is because the energy distributors often own the grid and have to provide an equal level of opportunity to all the suppliers, and from a technical point of view, it is unlikely that different PLC technologies will run simultaneously in the same physical environment. A solution could be that the distributor plays a role with data access. However, this can create other problems like adding limits for differentiation and competition, legal requirements, and dependence.

Specialized lawyers should be associated with smart metering projects to ensure that these implementations are carried out according to all legal requirements.

### Collaborators

People with different specializations and interests should actively participate in these projects so that all the branches of a utility can benefit. A lawyer does not necessarily understand smart metering systems. The same can apply to marketing professionals. All of these people should be provided with at least a basic level of technical information about the projects.

As in any automation process, utilities must deal with job security concerns. People sometimes believe that they or their colleagues will be laid off when an existing function like meter reading is eliminated. It is important that utilities adopt a transparent policy by properly reassigning people to new areas like monitoring centers, field marketing, meter installation, and safety inspections at customer premises. If necessary to make job reductions, this must be carried out in a way that minimizes trauma for the employees. Prioritizing reductions, for example, can be based on already-scheduled retirements. Efficient communication is again vital for this process.

## Conclusion

International benchmarking demonstrates the different needs of smart metering projects around the world. The time necessary to build these platforms and prepare the environment for this new scenario also varies. The different participants must actively work on these projects from conception. Customers play an important role in all this, as targets can only be achieved with their cooperation.

Drivers and opportunities surrounding these projects are innumerable, although they must be properly investigated and evaluated in order to help the different parties to extract the best value from each. Different aspects can be associated with each; for example, energy costs and prices can be potentially reduced, network management improvements are achievable, and demand in terms of energy consumption can be optimized in relation to available offers.

Rates offer opportunities to all participants. Complex structures and profiles can now be offered to consumers. Traditional views have also been upgraded; for example, a meter is now able to switch from debit to credit mode, and vice versa. Prepayment services can now become PAYG and enable even more opportunities.

International experience shows that these projects are likely to succeed, but this depends on the way they are designed, implemented, and updated during the life span of their components.

As with any other project, challenges must be faced. Systems must be future-proofed to enable business case assumptions to be effectively achieved, for example, in relation to life span in the field. Risks should be properly identified, mitigated, and monitored.

The next chapter discusses some actions utilities can take while developing solutions to reduce some of the risks and achieve success with deployments.

# 4

# Building the Technical Solution

The previous chapters discussed the need to reduce the risks of smart metering projects. These risks are mainly about the challenge of defining a technical solution. While the technical team is usually responsible for this part of the project, they should not work in isolation. Ensuring the strategic objectives of the company are met for the specified product requires the cooperation of other areas of the business. All teams report to the overall project leader and are all subject to company rules and methodologies. Thus, other tasks may be executed in parallel with the technical ones. Different workgroups are often created to deal with specific project management topics, such as defining strategies for marketing, finance, legal, communications, and regulation. This chapter is dedicated to the different aspects of building a technical solution, because of its core role (fig. 4–1). A bad choice at this stage could affect the strategy of the company and the benefits to be achieved.

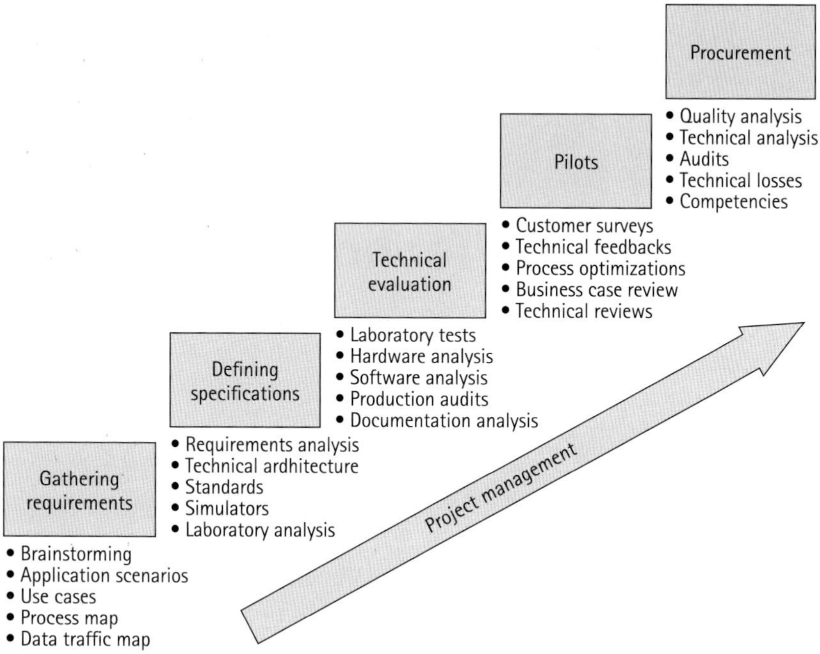


Fig. 4–1. Building a technical solution—stages diagram

# Gathering Requirements

Before defining the technical architecture, we must understand the requirements of the company and transform business needs into specifications. The technical team is essential for achieving this objective. It is generally composed of professionals from different areas of the business, such as research and development specialists (social, metering, IT, standards, generation, energy efficiency, and several other areas), field operations, corporate IT, and business specialists from different branches. The members of this team need a technical background as well as the ability to listen to the clients of the project. This is very important because the technical specifications must reflect the requirements of the business. Unfortunately, this is not always the case. Sometimes these teams try to impose predefined solutions available in the market, not necessarily compatible with the needs of the utilities. As we are dealing with a new environment, existing products may not deliver the differentiation levels required by companies to survive in a competitive market. A bad choice can limit the opportunities for the project. That is why specialists from different branches must be present and participate to provide the liaison with their own businesses and needs.

The process of gathering requirements is composed of different stages:

- Collect ideas from brainstorming sessions.
- Transform them into initial business requirements.
- Conduct requirements analysis.
- Refine them with others from international benchmarking exercises.
- Define application scenarios and products.
- Map and model business processes.
- Perform data traffic analysis.
- Manage documentation.

All of the assumptions and associated risks should be regularly logged during the entire project from its conception. The project team should conduct financial analyses of the risks and develop mitigation plans to address any problems that might occur.

## Brainstorming exercise

During this first process of gathering requirements, a fundamental tool is brainstorming (fig. 4–2). As the name suggests, experts from different areas of the business gather ideas collectively. This exercise collects important requirements for the project and motivates people, but it may produce a negative result if badly managed. For example, if someone's ideas were exposed to public criticism, this could end the person's motivation to participate or have worse consequences. Therefore, it is crucial that the exercises be led by professionals with extensive experience. That is why specialized consultants are often engaged for this purpose.

Brainstorming exercises involve people both within and outside the company. Several sessions are often necessary, and when these exercises involve customers, they should be separate from those attended by corporate and internal specialists.

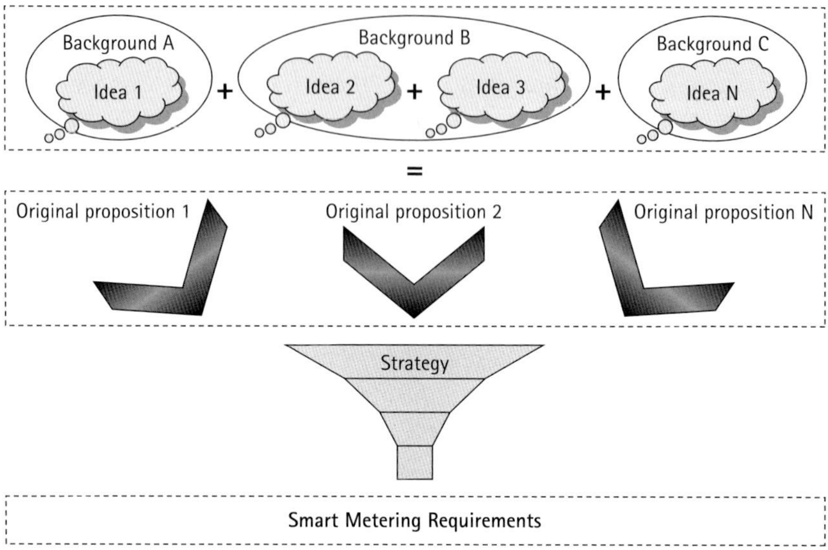


Fig. 4–2. Brainstorming exercise diagram

The following are some simple recommendations:

- Objectives. The goals must be fully defined to avoid nonproductive discussions.
- Environment. It is advisable to use an neutral location, preferably outside of the company premises.
- Time. Long exercises exhaust people and do not produce good results.
- No censoring or hierarchies. Everyone needs to be able to express their ideas with freedom. There are no ridiculous ideas and no bosses during these sessions.
- Organization. Everyone talking at the same time does not produce effective results.
- Take notes. All ideas must be fully explored and recorded.

Once collected, these ideas should be transformed into requirements. The team should group them, eliminate duplications and unachievable targets, and provide strategic analysis. This filtering process should be individually executed per requirement. Different aspects should be taken into account such as safety, finances, technology, future-proofing, and risk analysis. External views are also welcome; requirements from benchmarking exercises may be used to complete and refine the list. The result of this overall process will produce the initial draft of the business requirements. All the branches must then review and approve this document so it can used as guidance for the project. Note that these are not the technical requirements yet; they are intended to reflect the vision and the needs of the business.

## Mapping processes

The processes involved in smart metering projects should be fully identified and mapped. The requirements gathered should be taken into account during this exercise. It is common for these projects to produce modifications in the way utilities execute some tasks. The processes should be mapped, taking into account the way the tasks are currently executed and the necessary changes. External entities and relationships must also be considered.

Several methodologies can be used. Unified Modeling Language (UML) is a commonly used method. It is a graphical method for modeling processes and data treatment. Beyond being an efficient technical tool, UML produces user-friendly documentation, such as sequence diagrams, use cases, and actor maps. Because multidisciplinary teams make up smart metering projects, the use of UML is strongly recommended; at a minimum it should be used as a basis for the process modeling.

Use cases are employed for mapping processes. They group different, detailed processes that together define system behavior. Use cases may interact with each other for different purposes. Thus, they may be grouped as part of application scenarios (fig. 4–3). For example, a data collection use case may be used simultaneously for different applications such as displaying online load profile via web interface, conducting nontechnical losses analysis, and performing load simulations. Thus, application scenarios essentially reflect utility operational functions or products and services to be offered to consumers.

During this stage, new requirements can be identified and must be promptly gathered. The gathering exercise should be continuous during the project.

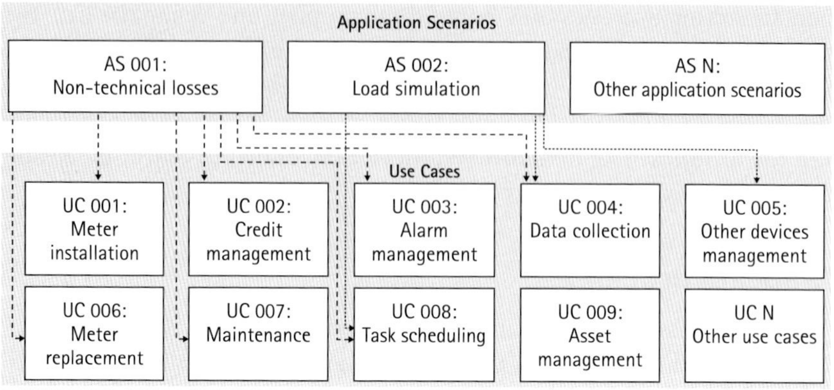


Fig. 4–3. An example of application scenarios

Application scenarios as well as their associated use cases and processes should now be transformed to data tasks and data transfer processes. This is essential to define some aspects such as communication media, topology, and processing distribution, as well as to develop service level agreements (SLAs). All processes related to data flows should be identified (fig. 4–4) and analyzed (fig. 4–5). Several related aspects should be documented:

- Data package attributes (e.g., task ID, meter number, date and time, and targeted parameter)
- Data package size in bytes
- Frequency of transactions (e.g., daily, monthly)
- Direction of traffic (e.g., upload or download)
- Origin and target of task (e.g., origin: middleware, target: gas meter)
- Priorities and SLAs
- Data profile (e.g., predominantly peak or off-peak time of communication for WAN)

After being identified and mapped, a data traffic study should be done to analyze several technical IT aspects, such as data storage, net data traffic needs, negotiation of eventual communication costs for WAN, and potential idle traffic to be negotiated in future with partners, if desired. This can also be used to refine and update the previous analysis.

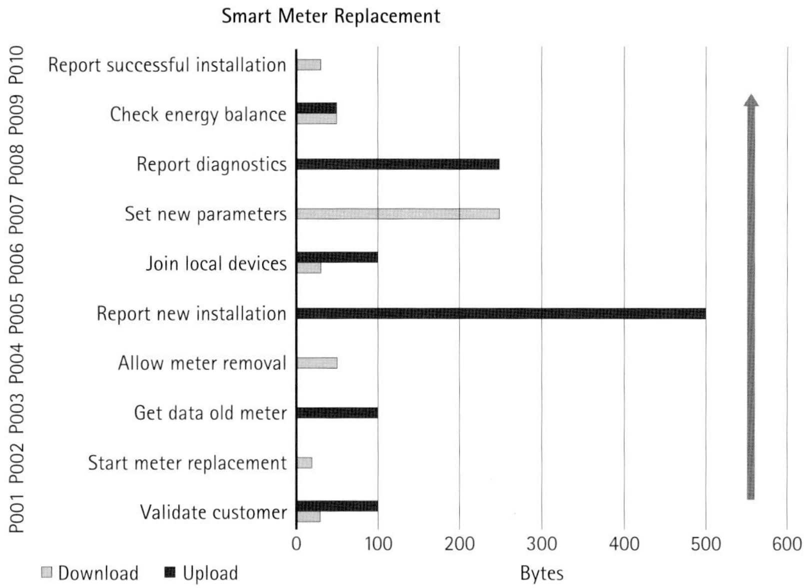


Fig. 4–4. A data traffic map

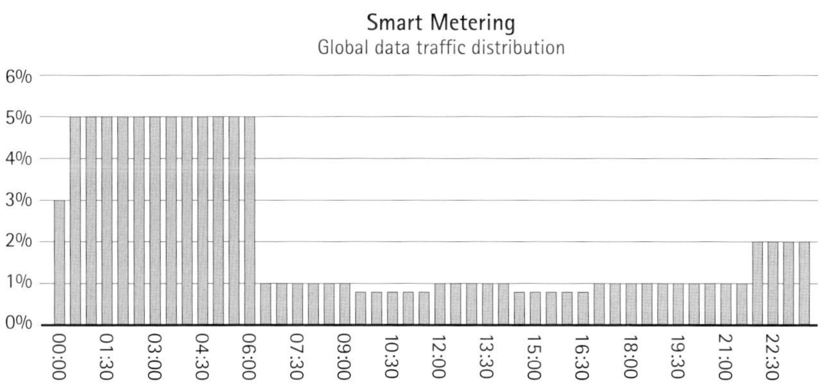


Fig. 4–5. Graphical data traffic analysis

All this detailed information should be documented; classified according to priorities and probability; and reanalyzed under financial, legal, safety, and other applicable criteria. After being reviewed by all the businesses, new versions of the requirements, strategic views, and product propositions should be created. They are essential for the next step: transforming the company needs into technical specifications.

# Defining Specifications

Now it is time to refine these requirements according to technical aspects and feasibility criteria. Efficient analysis is essential at this stage because, unfortunately, some products are not technically feasible. This can be due to several aspects such as safety recommendations and metrological rules.

The defined specifications will form the basis for building the appropriate technical architecture for the project. It is important to cross-reference the specs against existing standards and norms to identify synergies or new requirements, such as safety, metrology, data security, data exchange, and communications.

All the studies' criteria and analysis must be specified, including topologies; comparison of communication options for WAN, LAN (local area network), and HAN (home area network) functions; and distributed processing in the different components of the metering system.

Another point to be carefully determined is data security. For example, data flows should be classified as confidential, financial, private, restricted, etc. Interoperability and integration aspects should be defined, taking into account the strategic objectives of the company.

All of the initial analyses should be reviewed and updated with parameters obtained from the technical choices. For example, the net data traffic previously defined will be associated with protocol overheads, acknowledgment messages, repetition profiles, maintenance and supervisory controls, and data exchange formats. The business case should be revisited and updated with any relevant information.

This is an important period to start the validation process of all the assumptions. For example, all the assumptions related to the technical architecture must be proven. To meet this requirement, appropriate tools must be in place such as laboratory installations and IT simulators and emulators.

## Building simulators and emulators

During the process of building technical specifications, it is important to define criteria to compare the different technical options in terms of communications, IT, and architectural designs. Simulators are interesting applications to fulfill this need. They are usually IT applications designed to analyze scenarios that could happen in reality but are unlikely to be physically reproduced in a production environment. For example, an exception like the failure of package deliveries is a process that could be simulated. A communication problem could freeze a server or an in-home metering gateway due to the volume of accumulated tasks to be processed at the same time. Questions for other scenarios to be simulated include the following:

- What would be the impact of a general blackout if all the electricity meters generated outage alarms at the same time?
- How much data will need to be stored in the smart metering system over 20 years? How will the data storage be managed?
- What is the impact in terms of IT and communications if a decision is made to collect energy profile data from millions of meters every 10 minutes instead of once a day?
- In terms of distribution of processing, memory, and tasks over the smart metering system (e.g., servers, gateways, and meters), what is the most appropriate architecture for the type of services the utility would like to offer?

Simulation tools are also used for technical evaluation processes in order to compare propositions from different solution providers in relation to required levels of SLAs, data traffic, processing, real-time tasks, and many other features. When customized products are developed, manufacturers and solution providers use simulation tools to execute validations and corrections during the design process for hardware and software.

Many simulators are currently available in the market. They are able to execute simulations for network, communications, and other purposes. Examples of these tools are QualNet and Jade. Traditional programming languages such as Java-based platforms could also be used to develop personalized simulators for specific needs.

Emulators are other interesting tools for comparison. They are combinations of hardware and IT applications that can reproduce common situations in the field and are used to compare different solutions in an environment close to reality. They can be used to reproduce global scenarios combining physical samples and IT systems using statistical principles. For example, the network performance associated with sending load profile data from 100 meters to an IT system could be simulated with a sample of just 10 meters (fig. 4–6), related software, and laboratory emulators such as impedance and noise generators.

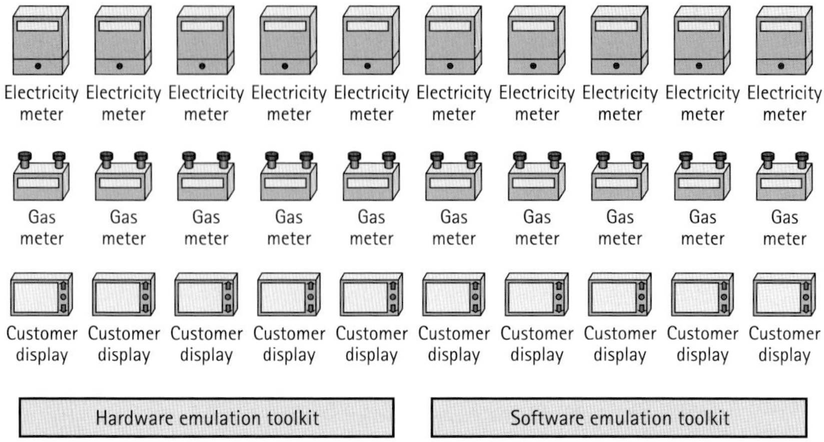


Fig. 4–6. An emulator tool

Utilities also use emulators to provide continuous feedback and tests of the final solution over its life span. For example, sample meters could be permanently installed in different environments, in a site owned by the company, such as their R&D premises. These devices should be exposed to normal field conditions like indoor and outdoor changes in temperature and usage. All the different meter classes could be installed in this test environment, from the first installation in the field. These devices should be controlled by the same middleware that manages the meters installed in the customer premises, although instead of being associated with its production mode, a special development environment built for test purposes would be used. After passing through the usual debugging and testing procedures, the different software versions should be exposed to this simulation environment.

The above environment can also be used to evaluate new applications under a real environment before they are downloaded to meters installed in the customers' premises. This procedure is essential, for example, for in-home metering gateways. As new market developments occur, new applications and related updates can be easily downloaded. Given the millions of combinations of conditions that these devices and firmware are exposed to in the field, it is unlikely all of them will be simulated in the laboratory environment. Thus, the use of emulators as additional test platforms improve the ability of identifying eventual bugs and faults for specific field conditions, prior to any deployment in the production environment. Among other risks, this can avoid irreversible damage to the image of the company.

All of these tools are fundamental for smart metering evaluations. Scenarios could be simulated and then emulated in order to identify and avoid problems that would be reproduced in the field. The findings from these procedures can result in changes of architectural designs, review of IT infrastructure, and redefinition of IT management methods.

## Laboratory tests

Beyond simulation and emulation tools, a modern and efficient laboratory is crucial for further testing. All solutions must be fully tested in a lab environment prior to installation in the field, even for small pilot programs. There are many aspects to be analyzed physically and logically in the solutions like safety and metrological tests, data security, and communication robustness evaluation.

Laboratories must be well maintained and certified by the relevant authorities. Professionals must be continuously trained. Metrological calibration, IT test sets, and other equipment must be properly managed. Labs must also be regularly upgraded in order to produce optimal results.

For smart metering projects, full environments should be provided in order to test all the different components, for example:

- Devices such as meters, customer display units, and gateways
- Communication evaluations (e.g., robustness to interference)
- Protocol analysis
- IT analysis tools (e.g., for data security and data management analysis)
- Interoperability and integration tests

Once the system is fully deployed, it must also be continuously evaluated. For example, samples from different meter families should be periodically removed from the field for metrological, safety, and other testing purposes (fig. 4–7).

Samples should be taken from the equipment purchased for laboratory evaluation prior to installation in the field. This is necessary for different reasons. For example, manufacturers could change electronic components inside the devices without previous consultation and evaluation by the utility company. One of the common drivers is reduction of production costs. Simple changes can have significant impacts on products, including their safety. Utilities must be attentive. Thus, laboratory premises and qualified professionals are vital for the success of smart metering projects.

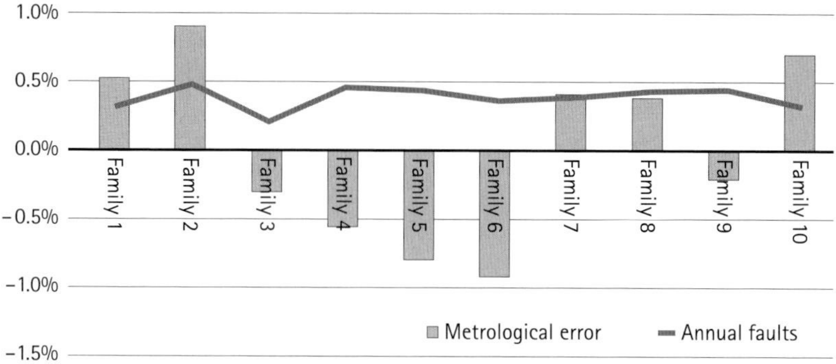


Fig. 4–7. A meter family analysis

Unfortunately, some utilities are halting these activities for different reasons and may pay an expensive price for it in the future. Due to the cost of the equipment necessary for testing (e.g., telecom, metrology, safety) and generally their infrequent use, utilities are finding it difficult to maintain their own laboratories. This is a fact, but it should not be used as an excuse to avoid fully executed tests. Equipment can be rented and services can be contracted for this purpose with specialized operators. Many laboratories are available in the market for such use. They provide turnkey evaluation on request or timely analysis, according to the utility's needs. Of course, all third-party services must be well supervised by energy utilities.

Expertise for executing these tests is not easily found. Energy utilities often do not have these kind of resources anymore, sometimes due to a strategy to focus on only their core business. However, consultants are available in the market to assist these companies, conduct tests, and provide reports. While the professionals must be carefully selected, if their competence matches the requirements and their contracts are well established, they can provide important feedback to energy utilities.

## Request for information (RFI)

Evaluation processes differ according to utility strategies and market rules. Once further investigation is approved by management, a formal RFI can be sent to different solution providers, Utilities should check the technical solution specified against the products available in the market. The results should be used to identify the next steps to be taken by the energy utilities and to validate time lines.

An evaluation process could also start in parallel. It often includes laboratory and field tests that will enable utilities to evaluate different solutions according to their specifications. Audits and other exercises should be undertaken to analyze the capability of a provider to develop new features, maintain schedules, respond to quality and sustainability rules, and deal with many other needs. Formal meetings are often required with meter manufactures to understand their products and to provide answers to their questions about the specifications.

Depending on the utility and country, it may be necessary at this stage to carry out certain legally mandated procedures such as publication of records. Regardless of the procedure to be followed, it is important to follow ethical principles and use transparent business practices. This is paramount to the success of any project.

# Technical Evaluation Processes

## Evaluation cycle

Deploying new smart metering systems can help utilities solve many problems, although it can cause other problems if the rollouts are badly executed. As mentioned, bugs and faults can damage the reputation of energy utilities. Tests are fundamental for avoiding problems during rollout. We discussed laboratories, which are important but are just one of the many tools that should be used. A full sequence of evaluations should be in place to ensure that products comply with specifications.

For hardware, prototype tests are usually employed. These are generally executed after the prequalification process is finished. Prior to full laboratory tests, which entail a great deal of time and cost, solutions should be evaluated and filtered using prototypes, with an analysis of the documentation needed to meet the requirements of a project. Examples of these parameters are capability of delivering technical solutions, certificates ownership, and other documentation analysis. Preselected solution providers are submitted to a full laboratory evaluation sequence. Products are usually evaluated in several phases (fig. 4–8):

- Prototype: In this early stage, solutions are initially submitted for evaluation tests. Final refinement of design and other nontechnical features cannot be done at this phase. Thus, the importance of this phase is to validate the technical concept, not the final product. Prototypes are normally used to provide technical feedback for procurement.

- Preproduction: These further investigations are normally executed for solutions chosen during procurement processes. They often incorporate changes from the previous versions and allow an evaluation procedure to be done under all conditions, including nontechnical ones.
- Production series: These are the final approved solutions. Additional tests are necessary to ensure that customization requested during the preproduction evaluation has been incorporated and that the required levels of quality are fully met by the industrialized sets of devices.

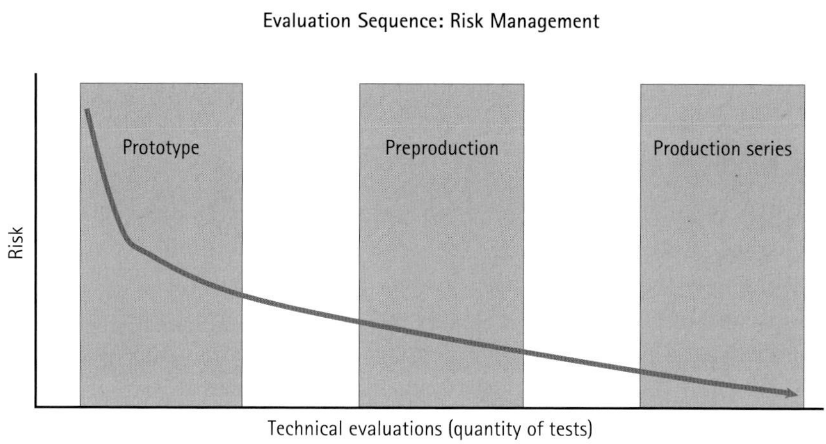


Fig. 4–8. An evaluation sequence

As observed, it is essential that tests be consecutively carried out to ensure that high-quality products that comply with specifications are deployed in the field. In terms of software, the procedure should be similarly executed. Databases are virtually populated, and events, like communication failures and outages, are generated to simulate daily operations and exception cases.

In both hardware and software cases, full test specifications must be efficiently prepared in order to understand all possible exceptions for smart metering systems before a rollout. Thus, they are used to ensure that the system is ready to be deployed.

## Hardware

Beyond metrological components, other aspects, such as load switches, valves, and communication devices are needed. Because of these new features, a series of tests needs to be planned. Standards should be updated to conform to this new environment and to the related legal and regulatory considerations.

A full range of tests should be provided for smart metering systems and their individual components (fig. 4–9). Here are some examples:

- Functional tests should be conducted at this stage, in which the main functions of equipment are verified according to specifications. Their behaviors within normal and abnormal environments are evaluated such as self-consumption profiles, starting current, switch operations, terminal screws resistance, modem operations, acknowledgments, rates, parameterizations, self-diagnostics, messages, alarms, and exceptions.

- Electromagnetic compatibility testing to evaluate impacts provoked by natural phenomenon or normal field conditions like lightning and radio-electric disturbance.

- Electrical influences testing related to the electrical environment like overvoltage, overcurrent, impact of network faults, external and internal short-circuits in the meter, and interferences naturally generated by the operation of medium-voltage switches on the network.

- Isolation tests to analyze different isolation levels in the meter such as between different phase terminals, local wired communication and phase connections, meter frames and terminals, and access to live parts of terminals.

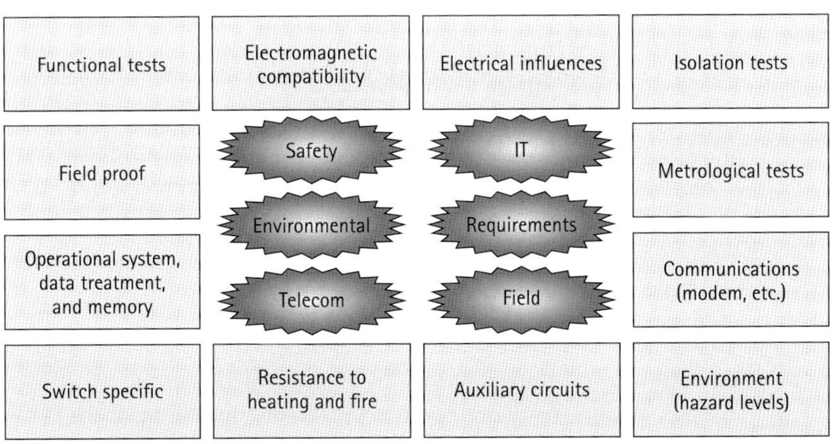


Fig. 4–9. A test diagram

- Field proofing tests of meters in the environment, which presents challenges depending on where meters are installed. Examples are humidity levels, heat and cold, ultraviolet effects, vibration, shaking, rain effects, and mechanical impacts.
- Resistance to heat and fire is tested, covering criteria such as self-extinguishment, flammability, and toxicity conditions.
- Environmental safety of components and materials is essential. Products such as batteries and electronic components should be less hazardous to the environment.
- Metrological accuracy tests are done under different environmental and electrical conditions, such as different levels of current, temperature, and humidity.
- Data treatment and memory are crucial for smart meters. Meter resets could be triggered, for example, to simulate the use of noise or electromagnetic-field-generation devices by ill-intentioned people. If registers are restarted, the meters could be vulnerable to nontechnical losses. In this case, the readings would return to the values registered during the last integration process. Thus, if they are registered once a day, up to 24 hours of consumption could be lost. It would then be preferable to use meters that are able to register these values in small windows, every minute, for example, and resistant to tamper. Data security is paramount, as already mentioned. Local data management is also essential to ensure security. Embedded gateway functions increase this need even more. Full tests should be done.
- Switch-specific devices entail specific tests. For example, it is necessary to ensure that the number of switch operations are compatible with meter life. Coordination with circuit breakers is also essential to ensure efficient safety functions. Short-circuit, overcurrent, and differential protections must be provided by circuit breakers without any interference by switch operations. Another point is customer acknowledgment for reconnection events. Except for outage events, it is unadvisable to reconnect the home without consumer awareness (acknowledgment).
- Auxiliary low-power circuits are sometimes present, as when meters use low-power contacts to control consumer load. It is thus important that these circuits have the required level of

protection, like fuses. They should also be fully isolated from the meter board and other circuits.

- Modem-specific equipment also requires special treatment. It can generate different levels of self-consumption in the meters depending on its status, such as standby or transmitting data. Batteries should be designed for ecological and user safety, especially when they are embedded in or attached to a gas meter.
- Operating systems should be tested. When an in-home metering gateway is embedded within an electricity meter, logic tests similar to those performed on computers are required. Supervisory modes and applications management are essential for future-proofing and require specific procedures of evaluation.

Beyond ensuring conformity to meter standards, these tests are crucial to guarantee the health and safety for users of these devices. They are mandatory for any smart metering deployment. Several other tests must also be executed, but these examples show the challenges in this new environment.

**Life span.** Many different procedures are used to accelerate the aging of devices and simulate their behavior in the field. They generally expose equipment to extreme environmental situations.

Discussions about the different methods of testing are so complex that they could easily fill another book, but tests made based on these different methodologies almost unanimously indicate two primary points of failure of smart meters: liquid crystal displays (LCDs) and electronic components.

Internal components of devices should be individually identified and evaluated by utilities. Their quality, life, and other features should be fully catalogued for evaluation and future comparison. Their circuits must be carefully understood to reduce risks associated with exceptions, such as short circuits occurring internally or on the meter terminals. Protection mechanisms must be in place to protect against short circuits.

LCDs can be even more carefully investigated. In general, when an electronic component fails, it stops operating completely, but the LCD sometimes fails only in certain places. If just one or two segments in a display matrix stop working, some numbers will not display correctly (fig. 4–10). These matrixes generally form an "8" frame, with different segments turned on to display different numbers. Assuming that the upper segment stops working, for example, it would be impossible to differentiate the number 1 from 7. In general, the only practical way to identify this anomaly is via customer feedback or when technicians visit consumers' homes.

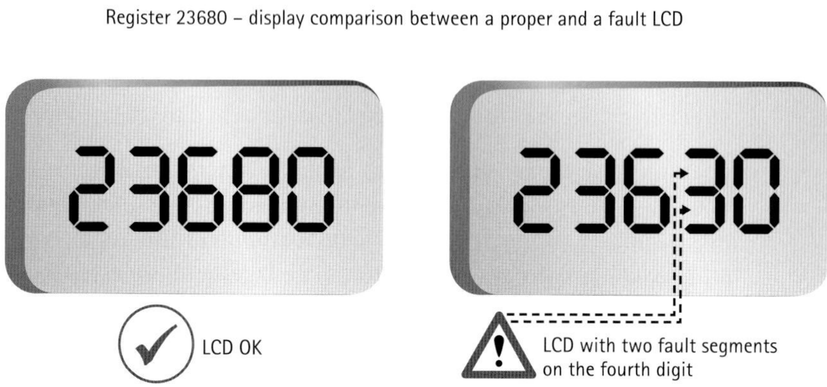


Fig. 4–10. An example of display failure

Many methods are employed to avoid this. A simple but efficient way is turning off LCDs when they are not in use. Assuming that customers interact with their meters only a few times a day, LCDs could be turned off more than 98% of the time in a sleep cycle, which could improve LCD life radically. Self-diagnosis could also be used to identify when display problems occur.

It is important to carry out aging tests. Component tests must be done to avoid premature meter replacements in the field, and preventive solutions must be planned. Methods are continuously improved to identify further vulnerable points to be treated and to improve meter life. This criterion significantly affects business case analysis. A simple improvement from 15 to 20 years on the life span of devices can make the difference between loss and profit.

**Production chain.** Beyond tests in utility laboratories, inspections of manufacturers' sites contribute to assuring quality of production. Audits at these sites are advisable in order to anticipate problems and identify risks. Random samples should be taken directly from the production line for testing procedures. Several criteria could be used during this evaluation process. Repeatability, for example, measures the quality of manufacturing for metering systems (fig. 4–11). Let us assume a meter specification that requires an accuracy class on the order of 2%. Assuming that 10 meters are taken from the production line and the percentage of errors is around

–1%, with variation ranging only from –0.9% to –1%, it can be assumed that these devices have almost the same behavior. However, a range of +1% to –1%, indicates that something could be wrong with the manufacturing process. Thus, at least a quality inspection would be recommended.

Another concern is interchangeable parts. There are many components that should produce the same behavior if a meter is replaced by another of the same model, such as load switches, terminal arrangements, and some electronic boards. If you cannot exchange them between a sample of meters, this could be a signal of poor quality and should be investigated.

Audits in situ are essential to avoid rejection of entire lots of devices and recall procedures after they have been delivered. Thus, identifying problems in the production lines can save costs for the different participants in smart metering projects, both for utilities and manufacturers. For this reason, these audits should not be seen as a negative point of view for manufacturers but as opportunities to avoid unnecessary costs such as shipping and rebuilding products.

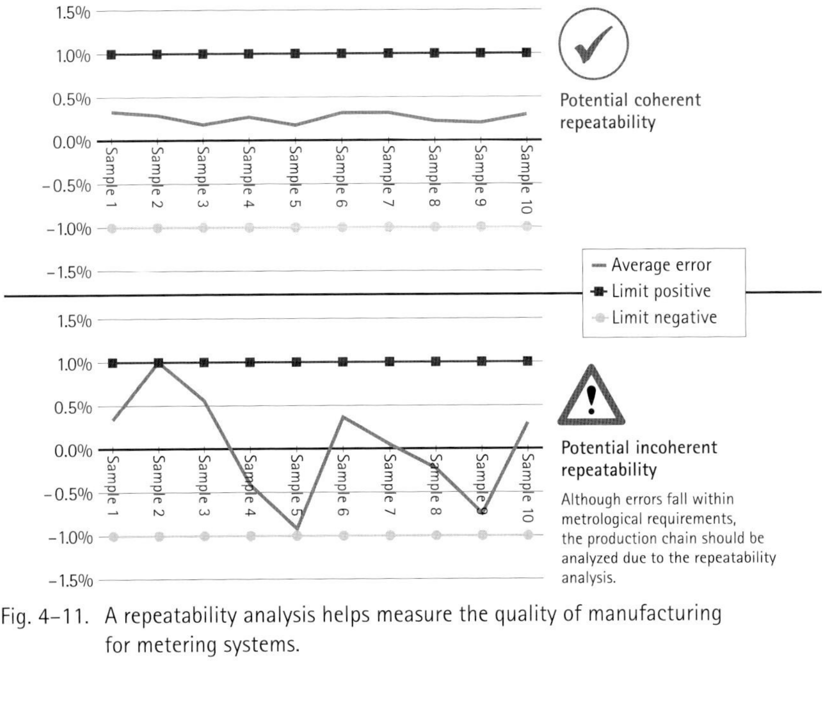


Fig. 4–11. A repeatability analysis helps measure the quality of manufacturing for metering systems.

## Communications

Most of procedures we have discussed for testing smart meters also apply to communications, although extra tests are required. As discussed, selecting communication media is not an easy task. A bad choice can restrict the services an energy utility could provide to its customers and opportunities for smart metering deployments. Thus, fully testing communication media is essential. The environment where these tests are executed and the infrastructure used for this purpose are also fundamental.

Each medium has its own specialties, and its evaluation process should take that into account. For example, with radio frequency (RF) communication, line of sight is a common reference to determine communication range. However, it does not take into account daily operational aspects such as reflection and absorption. A test using identical devices could produce different results if executed in different environments, such as near a river or in an open field. Thus, although this parameter is useful, it is unlikely to be used in isolation for media comparison purposes. Realistic field tests or emulations must be used.

With power line carrier (PLC), the situation is similar. Manufacturing specifications often say that PLC is able to communicate over ranges of miles without using repeaters. However, do these data really mean anything in isolation? In terms of real utilization, it is necessary to associate, for example, information related to impedance on the grid and attenuation of specific environments of an energy utility. Depending on these values, miles can easily become only tens of yards.

Another medium option is mobile communications. Coverage is usually the key point for selling these services (fig. 4–12). Telecom companies often claim great coverage levels over utility territories, although they usually refer to mobile phones. It is important to understand that mobile and fixed devices have different features and behaviors. For example, how many times have you tried to make a call from below a stairway or within a metal enclosure? In some countries, meters are often located in such places. If you cannot talk in the kitchen using your phone, you will probably move yourself to the living room if you know there will be better signal penetration. Meters are unlikely to be relocated for this purpose, and even so, as signal penetration can change over time due to different factors, several replacements would be required.

All other communication media have similar problems. Thus using laboratory testing is essential to assure compliance with norms and standards and to provide required simulations and emulation of different situations that could be found in the field. For example, different sources of disturbance should be generated in order to evaluate the behavior of the solution under interference. Of course, health-related tests must also be run to ensure that there will be no effect on humans.

Finally, testing communications media requires experience, the appropriate technology, and a high level of attention in order to apply relevant criteria.

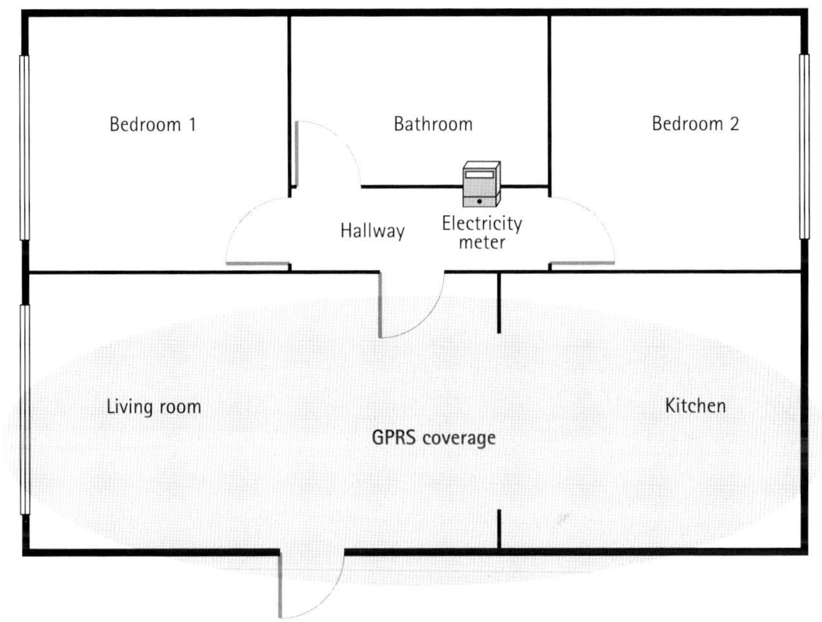


Fig. 4–12. Mobile coverage within a home can have blind spots.

## Systems

Different IT systems are used to support smart metering platforms like operating systems and applications for in-home metering gateways, firmware for devices, and modular software and agents for middleware. Based on this, in order to deal with this technological world, a test environment is essential for certifying versions before their effective use (fig. 4–13). In addition to the usual tests executed by solution providers and program developers, simulation and emulation tools should be developed for test purposes, as previously discussed. Scenarios composed of a large number of rules should be created to analyze all the functions, behaviors, parameters, and exceptions for every version of purchased products. In this way, it is possible to confirm that no regressive or collateral effects were generated from one version to another.

Test scenarios are also necessary because the main opportunities generated by smart metering systems are for new applications such as rates, display information, and integrations. Downloading a new application to a large number of in-home metering gateways without full test procedures could cause many future problems for utilities, including damage to the company's reputation and increased costs for support and maintenance. Significant costs can apply due to the increased number of visits in the field to reset units if conflicts cannot be managed remotely.

Smart metering systems are essentially composed of logical components, and they must be adequately tested. If the company has no specialists to deal with these components, it will need to hire or contract with outside professionals. Alternatively, the scope of the project would have to be reduced to a simpler environment, although this would compromise the opportunities related to these platforms.

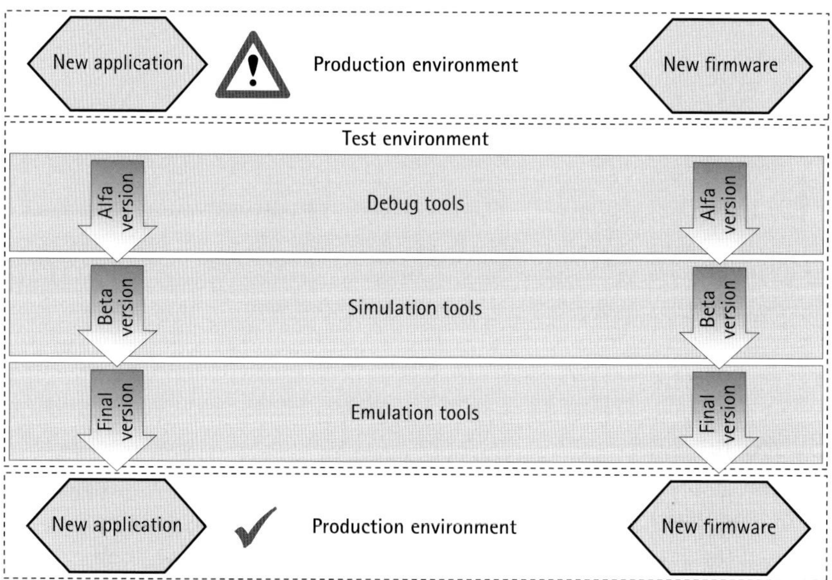


Fig. 4–13. The process of new application deployment is created in a test environment.

# Pilot Programs

Field pilot programs are usually the step that immediately follows laboratory tests. They are essential to any massive installation in customer premises. Among their advantages, they can gauge the reaction of consumers to new technologies. Sometimes, a technology suitable for engineers and marketing specialists is not necessarily adapted to customer needs. Thus, this step is a listening opportunity. Feedback provided by customers is essential to refine smart metering products and to produce solution corrections to designs and specifications.

There are also opportunities to evaluate the challenges for deployments in relation to installation, maintenance, and daily operations. Assumptions should be confirmed and extra recommendations produced to improve field operations.

Initial technical feedback of experiences and benefit-realization exercises are often made at this stage to confirm assumptions about these deployments. Cost estimations and risk logs should now be reviewed, as well as lists of opportunities.

More than just a learning exercise for all the different participants, it is a key moment of decision for the project. Pilot programs can halt implementations because of poor business model reviews, but can also identify new opportunities not previously seen. The next sections study these opportunities.

## Technical architecture

In addition to opportunities for defining the best products to be offered and ways to interact with customers, pilots are also an opportunity to challenge the technical architecture of a project.

It is recommended that all the potential communication technologies for WAN, LAN, and HAN be tested in the field. Results of tests in the field can differ from those executed in a laboratory environment. Thus, trials should be provided to ensure that practical and theoretical analyses are aligned (fig. 4–14). For example, a material can have different absorption features for RF waves, depending on its age and condition. A wet or humid wall produces a different effect for RF emissions than a dry one. Sometimes, it is not possible for energy utilities to simulate and emulate all the field possibilities in the laboratory, due to cost effectiveness and feasibility. Natural phenomena are very hard to reproduce in a lab. Rain and snow events can significantly affect RF transmissions and can often only be fully evaluated in pilots in the field.

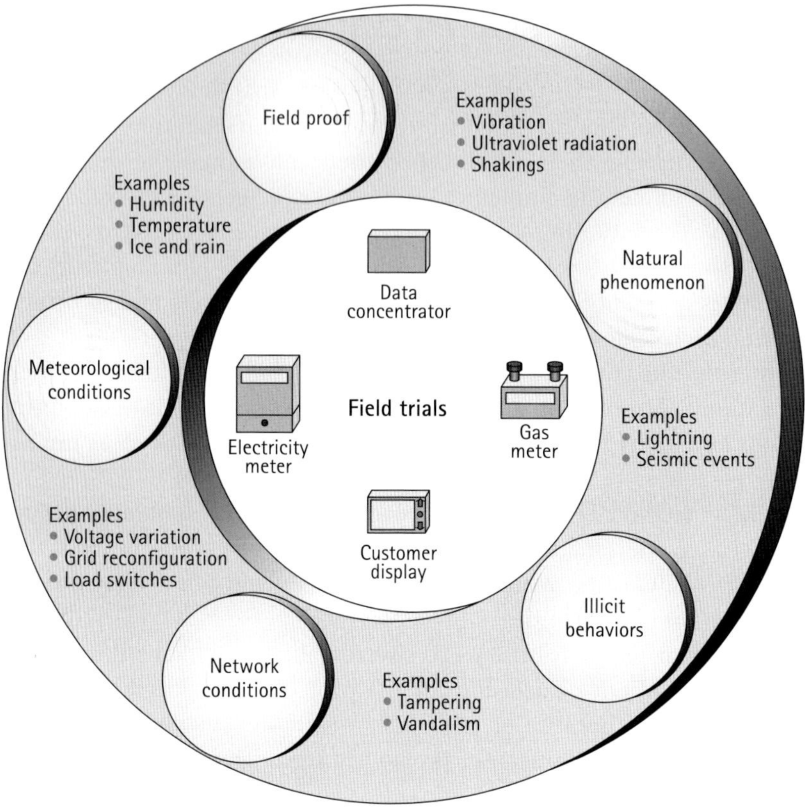


Fig. 4–14. A diagram illustrating environmental analysis during field trials

It is therefore important that all situation changes and effects be recorded during these tests. While smart metering devices were once commonly used as the only measurement devices in these evaluation processes, this method has proven to be ineffective. Several other devices should be used to fully map the field environment where these tests are carried out. Conditions in the natural environment cannot be controlled for as in a laboratory. Temperature and humidity cannot be controlled in the field. Any occurrences such as variation in weather conditions, network quality, seismic events, network reconfigurations, and physical damage caused by events like vehicle collisions with electricity cable boxes or posts should be monitored in detail, at least for an initial period. These data should be cross-referenced with smart metering data to identify any collateral effects they can produce.

The number of devices used during a trial is also important to produce useful forecasts, which should be defined through statistical criteria. An expert in this area should participate in designing the field trials. For example, a test of a dozen meter points is unlikely to statistically represent one million customers. The quantity should be at least enough to produce the corporate changes required for this new technology, enable a full review of business case assumptions, and engage the organization from the start of the project.

A functional test is also essential. Even if some functions are not naturally reproduced, they should be tested to emulate an occurrence. Remote connection and disconnection, for example, cannot be reproduced naturally for a customer segment. Legal arrangements should be in place with customers to enable all the functions to be fully reproduced in the field, with as many exceptions as possible. Planned, brief communication failures could be reproduced to confirm simulation and emulation assumptions. It is better to produce these events at this stage than to deal with these problems later with millions of consumers. Some companies use their own sites in these pilots, because of legal and ethical concerns, to avoid conducting emulations in customers' premises.

Finally, no architecture should be defined without full tests in the field. Tests are the only accurate way of confirming assumptions and technology promises and to emulate the corporate changes required for the deployment of these projects.

## Reaction to products

Different opportunities are afforded by smart metering systems. The majority are for new products and services offered in a specific market. Demand-side management is an example of an innovative service. It is a very important tool to achieve energy efficiency targets, but it needs to be fully defined according to consumer and utility needs. Affects on the comfort of customers should be minimized, so, it is very important to learn lessons from pilot programs that will contribute to improvement.

While solutions must be flexible, providing too many options can confuse rather than help customers. Pilot programs are a good opportunity to group them in efficiently targeted options. As it is a learning exercise, at least in the beginning, regular exchanges with customers are encouraged to ensure that energy consumption behavior is changing. For example, customers might decide that standby loads should be switched on every day at 6:30 a.m., but then realize that on weekends this time should be 8:30 a.m. They could also experience difficulty reconnecting the device when their schedules require a change from the regular programmed time. In addition, they may need to either raise or lower the thermostat setting.

Customers may need advice about new rates. Color codes associated with rates may be confusing at first. If a rate is for demand-side management offers, customers may require advice from energy utilities regarding priorities for their different loads and for power limitation purposes during peak hours.

Generally, once the initial settings are made, everything tends to be easier, automatically managed, and transparent to customers. However, due to problems during initial settings, customers can resist using these technologies. It is then vital that energy utility staff be available for consumers during this period. Attentive listening and prompt correction are very important. Feedback with no correction or full answer can discourage people from actively participating in the process. Efficient communication is fundamental now. Consumers must trust this new technology and the energy utilities, and realize that together both consumers and utilities can achieve their targets. However, achieving this level of confidence requires hard work, dedication, and patience.

On the utility side, it is also important to reduce risks in business models. Among other advantages, smart metering platforms enable utilities to offer aggressive energy-efficiency propositions for their consumers (fig. 4–15). Sometimes these products create risk in contracts. They are offered to residential customers but also to medium and large businesses. Thus, the financial risk varies widely. For example, some energy utilities currently offer to finance the deployment of specific energy-efficiency solutions without any starting cost for their customers. To understand this option, a case study follows.

Assuming a supermarket chain would like this service, the utility could send energy-efficiency specialists to their premises. The utility carefully studies the supermarket chain's installations, consumption habits, and other parameters to decide the best offer to propose. The utility can also install advanced submetering devices and devices that measure energy quality to make a detailed analyses of consumption per subsegment (e.g., bakery, offices). At the end of the process, business case studies would be prepared to study different energy-efficiency options. At this point, the customer could be offered a price reduction of $400,000 for the next year based on current consumption levels. All the required equipment to enable these savings would be installed on a turnkey basis at no initial expense to the customer. An annual payment of $250,000 would be required at the end of year one, but only if the target savings was achieved at this point.

Similar services can also be proposed to residential customers via intelligent algorithms and tools, analyzing the data generated by smart meters compared against other external databases.

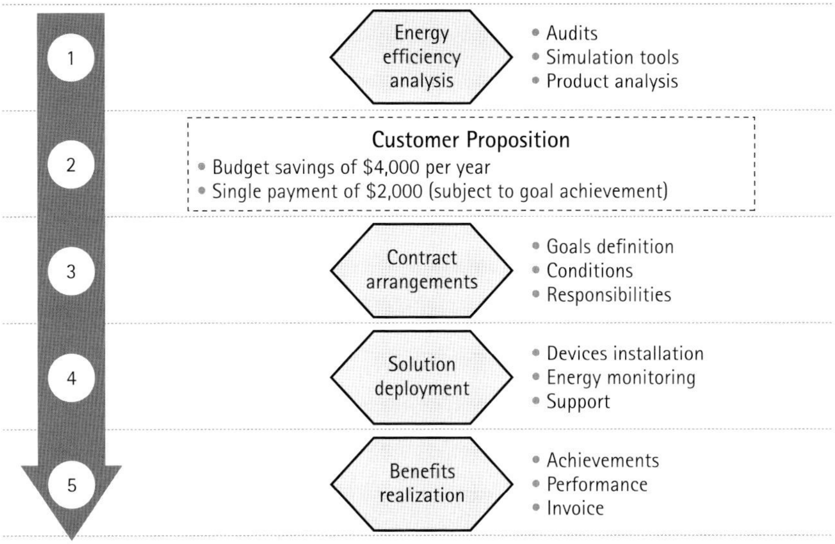


Fig. 4–15. Aggressive energy-efficiency proposals

The opportunities for these proposals are very interesting, not only financially but also in terms of reputation, marketing, building the trust of customers, and other indirect aspects. However, the risks are quite high. Thus, companies should be sure about their ability to achieve results before proposing these services. Pilot programs are essential to test, develop, and mature proposals. Products could even be offered for free as a way to test assumptions and methodologies.

## Communication channels

Communicating with customers is a core requirement of smart metering projects. If the customers interact directly or indirectly with the utility through a given channel, communications must be defined according to their needs. Pilot programs are good opportunities to test concepts.

**Payment methods.** For services like pay as you go (PAYG), it is important to ensure that customers have the most efficient payment options. There are many technological alternatives for consumers to make payments. They can be as high-tech as using short message service (SMS) media for this purpose or traditional methods where cash or credit transactions take place in pay points such as partners' stores. Hybrid options could be tested during this period.

Due to cost, often an automatic and electronic method is preferred for energy utilities rather than local payment with cash. Reduction of security and financial management costs are generally strong reasons, among others. Not only should business model analysis be considered, but also consumer demand. Some people are not still confident with modern technology. For others, such as those on fixed incomes or people who normally receive their payments in cash and do not have a bank account, payment in cash is the only choice. Therefore, the implementation of automatic support technologies is necessary, but different customer needs must be taken into account.

It is common to focus attention on the payment interaction itself and forget about other media. In addition to the payment method, there is another important communication channel—the receipt. It is sometimes treated as just a piece of paper, but it is a core report that can help avoid unnecessary costs. A large share of call center contacts is caused by misunderstandings, often about complex and low-value billing reports. Thus, receipts must be carefully designed to provide all information in a user-friendly and understandable manner. This is even more important when receipts are used to demonstrate other transactions, including the payments of debts and non-energy-related services. Thus, different receipt format options should be part of the pilot program's scope.

**Display interfaces.** There are many display options available, and some can be more suitable for particular consumers than others. For example, students may prefer to use mobile SMS, while stay-at-home parents might prefer in-home display panels or televisions. Pilot programs are good opportunities to test such assumptions.

Different technologies should be available for customers in the same market segment to test assumptions. Utilities should also try to use existing devices such as televisions as customer display units to avoid unnecessary consumption from customized devices. However, if utility-customized display panels are identified as essential during these pilots, these panels should not add new power consumption to the homes, or they should use the lowest possible amount of power. Environmental impact must also be investigated. Thus, energy utilities should make sure customers are aware of these issues and offer options. For example, some utilities are deploying solar-powered in-home display units supplied by embedded photovoltaic cells and batteries, which requires no extra energy for operation.

Several parameters can be evaluated (fig. 4–16) such as user-friendly design. These interfaces are commonly able to log the number of customer interactions, provide comparisons, and evaluate media acceptance. Parallel to this process, efforts should be made to quickly gather user feedback.

Customer privacy must be ensured. Products that are considered invasive during these exercises must be reviewed and corrected to avoid future rejection. For example, too many alerts can be seen by customers as commands rather than helpful advice.

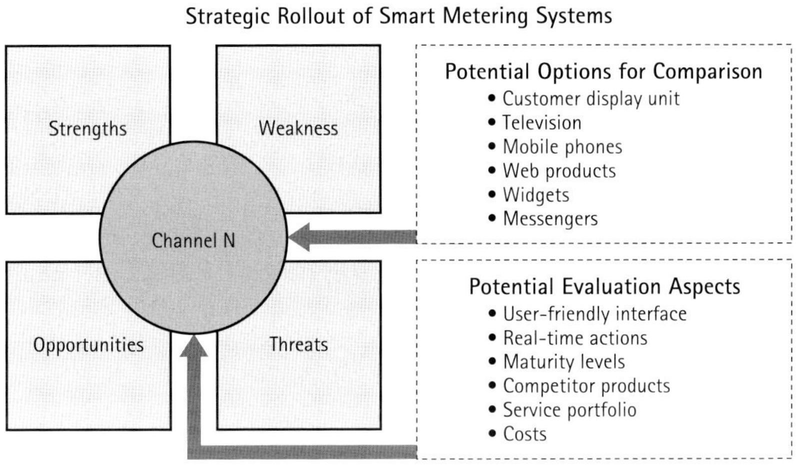


Fig. 4–16. A channel evaluation comparison

Training and education campaigns should help consumers deal with new services and technologies. The full potential of these tools should be used. For illustration, many people now have modern "smart" microwaves and washing machines in their homes. However, how many use even 20% of these smart appliances' capability? This is generally a result of bad communication, in which customers have to learn about these technologies on their own with no support.

**Other methods of communication.** Smart metering projects enable the use of different methods for energy utilities to communicate with customers. There are a large number of options, some of which we have already studied. This section explores a few others.

One of these methods is through reports. Many different types of reports can be tested in this phase, as well as the way they are delivered to consumers, such as by correspondence or face-to-face interactions. Electronic methods are preferable to paper communication with respect to environmental friendliness.

There are many types of reports commonly used, including:

- Billing invoices
- Press communications
- Marketing messages
- Appointment tools
- Interactive applications
- User guides
- Brochures

The communications package should be carefully built and tested. Unfortunately, many communication methods fail because customers either do not use the products at all or use them for only a short period. Some questions could be posed: How many people fully read user guides? Assuming they read these guides, do they fully understand them? How many marketing letters do people read before discarding them? Careful market research is required to maximize efficient communication choices.

**Surveys.** Surveys can be efficient way to gather customer feedback and providing necessary corrections in products and services. The first aspects to be defined are what questions need to be asked of consumers. Although new questions can appear, as a consequence of field deployments, at least a minimum scope should be established prior to beginning trials in the field. Thus, a list of questions should be carefully prepared and validated beforehand. Sometimes, poorly formulated questions can produce erroneous results and can result in extra costs and additional research.

Methodologies need to be established too. There are two main types of surveys (fig. 4–17):

- Quantitative surveys are often based on statistical monitoring parameters and are converted into data for companies. They can be targeted to a large number of people. Examples are Internet and telephone quality-of-service monitoring.
- Qualitative surveys are much deeper and are carried out for just a few targeted people. They have detailed questions and provide voluntary feedback. They can be used to validate and refine questions from quantitative surveys.

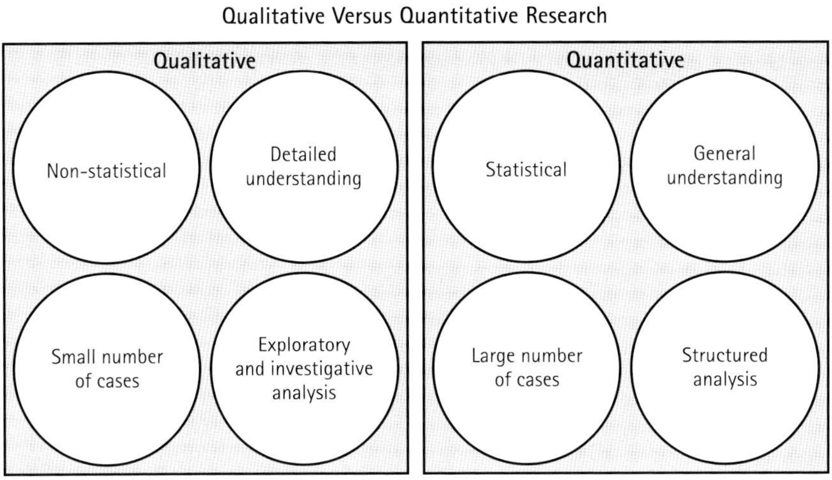


Fig. 4–17. Quantitative versus qualitative surveys

The most adaptable methods should be defined depending on needs. For example, questions related to display interfaces could require detailed feedback (qualitative analysis), while perception of reduced consumption after the project is deployed can use direct questions (quantitative analysis). Analysis of results is a crucial point for understanding feedback and ensures that the correct approaches were in place.

Due to complexity, specific expertise is needed to elaborate and execute these exercises. It is also important that results are completely impartial. Based on this, external specialized companies are usually hired.

## Business model reviews

Pilot periods should also be used to validate business model exercises in full, when possible. For example, some planned opportunities cannot be achieved if customers do not use a specific product or do not see any added benefits in paying for a specific advanced service. Financial analysis should be based on field proven exercises, not just theoretical ones.

Benefit realization analysis should start during the pilot period rather than after massive rollouts (fig. 4–18). This anticipation of results can provide market participants with a more robust concept of the achievements to be obtained and related cost adjustments. This methodology reduces the risk of investing large amounts of money purely on theoretical analysis. It reduces the risk of unrecoverable investments and can trigger strategic reviews such as an early halt to the project or financial share redistribution among different products and services within a project. For example, investment in demand-side management products can be more beneficial overall than in-home security services. In this case, at least part of the budget initially attributed to one product could be reallocated to another one.

Field pilots should be fully designed to provide all necessary feedback for the financial team. Related exercises are only possible if the fundamental parameters are included in these experiments. Strategic review must be a continuous process.

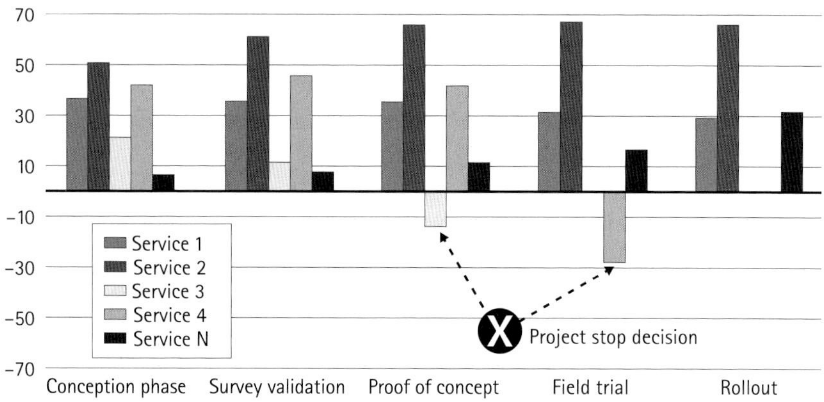


Fig. 4–18. Benefits realization during different project phases

# Procurement Analysis

At first glance, procurement could be seen as a purely commercial exercise. For smart metering projects, this assumption has been proven incorrect. Technical evaluations are important elements for decision making and should be used as one of the main criteria to rate vendor bids during procurement.

When technical evaluations are only provided to filter potential solutions, products and services tend to be purchased according to price criterion only, without taking into account important aspects such as quality and indirect costs that deployments can also generate.

## Quality analysis

Quality of products can affect deployment of technologies, generate costs for replacements, and impact company image. These factors should be fully measured and transformed into costs to be part of the financial analysis of a product. Comparison must be done under a common basis. To better understand this aspect, let us assume that a specification establishes a minimum of 15 years for a device life span. During laboratory tests, if two devices comply with the specification, then they would be approved using this criterion. However, as these devices are not equal, they have different behaviors. The first device can last for exactly 15 years, whereas the second may have 25 years of estimated product life. In this case, their costs are completely different for the energy utility. The additional 10 years significantly change business case analysis because of the product replacement cost of the 15-year product in the field.

Therefore, the price of a product is not only composed of financial value of acquisition but also to the costs related to its full life, from its conception to its removal from the field. For further analysis, let us assume another scenario. Battery life is fundamental for gas meters. One product can initially be more expensive than another, but if the more expensive product has double the battery life as the cheaper product, its total price could be lower. In this case, it would be advisable to buy the more expensive product instead of the initially cheaper one.

Meter fault levels are another interesting quality criterion. It would be unfair for a device with annual reliability values of around 0.5% to be treated the same way as another with 1%, even if the base requirement is 1%. If manufacturers focus on aggressive targets, this should be taken into account during procurement process.

Accuracy is another point. A meter that can ensure ±0.2% of metrological error cannot be compared with one that is offering ±1%.

Note that most technical aspects are likely to be transformed into financial values. Thus, they should be considered in procurement processes. Of course it is also important that rules be well defined for all the participants in a procurement exercise. Technical and procurement specialists should work together to define compatible test methodologies, evaluation tools, conversion parameters, contract arrangements, and other related factors to ensure that an efficient, sustainable, ethical, and transparent process is implemented.

Contract arrangements should be audited with continuous reviews to guarantee that all the claimed costs are being maintained. For example, fault levels can be defined and any excess above a defined limit converted in a compensation fine commensurate with the impact on the business. In this case, an extra penalty could be in place for serious nonperformance. Another example is that testing should be done at various stages of equipment delivery in order to detect any variations in quality. Similar fines should apply here. These procedures are necessary to ensure that production devices are delivered with the same features as those initially tested. A combination of procedures is needed for comprehensive testing.

## Technical losses

Another criterion is the power consumption of equipment when installed in the field and therefore connected to the grid. As different products have different levels of consumption, a measure of their electricity use should be part of the financial consideration.

As an example, take two specifications for a common electricity meter with a maximum burden of 1.5 W per measurement element (fig. 4–19). Let us assume the price of Product 1 is 5% lower than Product 2; let us assume Product 2 costs $30. We could initially conclude Product 1 is more advantageous, but is it really? If Product 1 consumes 1.3 W and Product 2 consumes 0.8 W, we would not be so sure anymore. In this case, Product 2 will consume 0.5 W less than Product 1. Assuming 20 years of meter life, this difference could represent around 86 kWh. Taking into account that a utility could be selling this energy in the residential market for $0.12 per kWh (e.g., without considering the capital costs over the period and other advanced finance analysis), we can easily gain more than 30% of the potential acquisition price of a smart meter device with high volumes. If added to the acquisition price, it is clear that Product 2 is a better return on investment than Product 1. A mistake resulting in the acquisition of Product 1 could cause serious problems for business models considering the millions of meters to be deployed.

Penalties should be in place for noncompliance with contracts. For example, assuming that a sample of meter prototypes were evaluated with 0.9 W on average but during the delivery of the production devices, they were found with 1.2 W, an audit process must start. If the meters were not immediately replaced, the costs of the consumption of 0.3 W for all the products, taking into account their expectation life, should be converted into a compensation fine. This cost should also be added to a penalty fine to be paid by the manufacturer. Investigation procedures are also recommended to understand the root cause of the problem. Depending on the conclusions, these could result in additional penalties or sanctions to the solution provider.

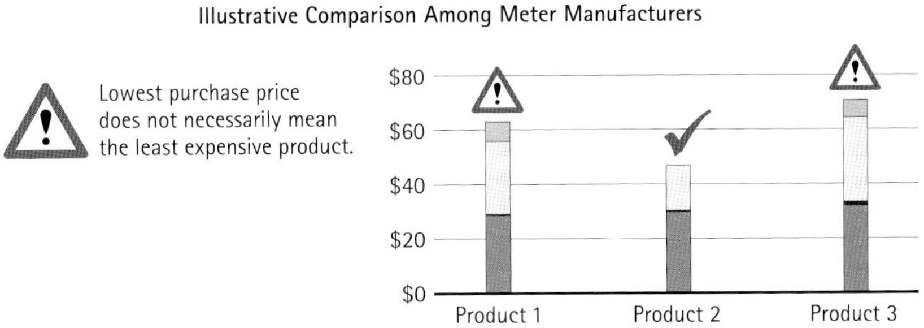


| | Product 1 | Product 2 | Product 3 |
|---|---|---|---|
| Meter life (earlier replacement) | $7.00 | $0.00 | $7.00 |
| Self consumption costs | $26.95 | $16.58 | $31.00 |
| Meter fault costs | $0.70 | $0.35 | $1.05 |
| Meter acquisition costs | $28.50 | $30.00 | $32.00 |

Fig. 4–19. A diagram illustrating the use of total price method instead of acquisition price only

Different profiles of consumption should be considered for smart meters, such as when they are on standby or communication mode, and when communicating under repetitive interference situations. Scenarios should be developed to make this analysis as accurate as possible. These scenarios could also be used for other purposes, such as to measure the battery life of gas meters. Note that this criterion also applies to auxiliary devices such as instrument transformers, modems, and gateways.

The customers should also obtain extra benefits from these evaluations. Assuming battery-powered in-home display units with different levels of consumption, the lower consumption would impact the customer less in terms of battery acquisition (assuming that the customers are responsible for battery replacements per legal agreements). However, the energy supplier's image of being responsible could also be affected, or it could be responsible for providing the batteries for replacements. A display unit supplied by photovoltaic cells could be more advantageous if compared overall to units containing replaceable batteries.

## Smart procurement

It seems clear then that smart metering projects require new methods of procurement. The acquisition price of metering products should be associated with operational costs to enable an overall analysis for procurement. This new environment requires changes to traditional approaches and different teams working together to maximize the benefits for energy utilities. Prior to smart metering systems being deployed, the same procedures could equally apply to the purchase of traditional electronic and electromechanical meters to reduce costs for utilities.

## Conclusion

Well-defined requirements are essential to the success of any smart metering project. The requirements-gathering stage can be a challenge for utilities. Due to the complexity, specialists must carefully undertake this. After filtering and analysis, this data will form the basis for business models.

When gathered, requirements need to be converted into technical specifications. They establish the initial basis for technical evaluation and procurement to take place.

A modern laboratory infrastructure is absolutely necessary to ensure that the specifications are fully met by the products developed and delivered by solution providers. Continuously trained specialists who can assist in the choice of the right technical solution to be deployed are also essential to maximizing the opportunities for all market participants.

Performing pilot programs in the field is critical before making mass rollout decisions. Different solutions must be evaluated according to technical and customer-related aspects. Strategy can be confirmed or changed during this stage, depending on the results obtained during these trials. Listening to the customers and providing necessary corrections are key to building the right products to be offered by a utility.

Smart procurement is essential. Operational aspects related to the life of the products in the field, such as the consumption of energy meters during their life span, should be converted into costs. They must then be added to the acquisition price of metering devices for full comparison purposes during procurement exercises. This is important to understand the real total costs of these products. Thus, a more expensive device in terms of purchase price is not necessarily less attractive in overall financial terms. For example, technical advantages could compensate or even outweigh financial differences. Among others, these products could have an extended life span, require less energy consumption, and be of higher quality, thus requiring less maintenance costs.

The requirements, specifications, business cases, and strategy must be continuously reviewed during the entire project life to produce refinements and corrections. This is crucial to ensure that these platforms are fully deployed and all their opportunities and targets are effectively achieved. The project team, using a structured project management methodology, should effectively manage the exercises studied in this chapter, as well as other analyses. Managing smart metering projects is the next subject covered in this book.

# 5 International Trends

Customers are becoming more sophisticated and energy markets more competitive. These require innovation of the products and services that energy utilities intend to offer their customers. Utilities are not limited to their core business; instead, they continue to extend into other areas according to their needs and as technology evolves. As a natural consequence, win-win alliances may be developed, targeted to both business-to-business (B2B) and business-to-client or -consumer (B2C) strategies.

A smart metering system offers numerous opportunities because of its capability for expansion of the scope of services and products energy utilities can offer their customers. As discussed, this is possible thanks to the ability to integrate with other external platforms. Many platforms may interoperate with these systems now and in the future. A smart metering platform is usually installed for more than 15 years. Thus, to maximize its benefits, a utility should comply with the short-, medium-, and long-term strategies of different market participants. Although some of these aspects have already been superficially discussed, this chapter expands on the trends related to these systems and explores further opportunities to integrate with other systems.

# Streetlight Platforms

Public lighting is a problem that challenges many energy utilities around the world. Although managing this service may appear basic, it is far from an easy task. The service providers responsible for public lighting face many challenges, including:

- Vandalism
- Flickering bulbs
- Bad contacts producing lamp sparking
- Faulty lamps
- Failures of photoelectric receptors
- Ballast failures
- Life span and preventive maintenance management
- Consumption monitoring

Even though privatization or decentralization decisions have shifted the responsibility for this service away from energy utilities, it continues to generate problems for these companies. Thus, many challenges are faced by these enterprises, such as:

- Difficulties of maintaining an efficient asset database (e.g., quantity and location of lighting assets)
- Uncontrolled replacement of lamps and associated devices that require higher energy consumption
- Clandestine connections within networks
- Technical losses on dedicated circuits and transformers
- Lighting turned on during the day
- Uncontrolled power factor of lighting ballasts
- Changes of consumption due to seasonal amount of daylight (e.g., in some seasons, lamps stay on longer than in others)

As you probably now realize, although streetlights may generate some inconvenience to utilities, they can also generate many opportunities of partnerships between energy utilities and lighting service providers.

Smart metering systems may enable these companies to work together for common benefits.

Additionally, benefits for both companies can go even further. For example, as these assets are everywhere, they may be strategic for smart metering deployments. Streetlight electrical boxes may be strategic points for the installation of low-power radio frequency (RF) repeaters to build a local area network (LAN) environment (fig. 5–1). Thus, these nodes can be used as repeat signals for many communication media. Sometimes the cost of repeaters significantly affects smart metering business cases. For example, it can be difficult to find places to install RF technology devices, and even when found, leases must be negotiated and paid. Streetlights offer perfect opportunities because of their height and distribution along streets.

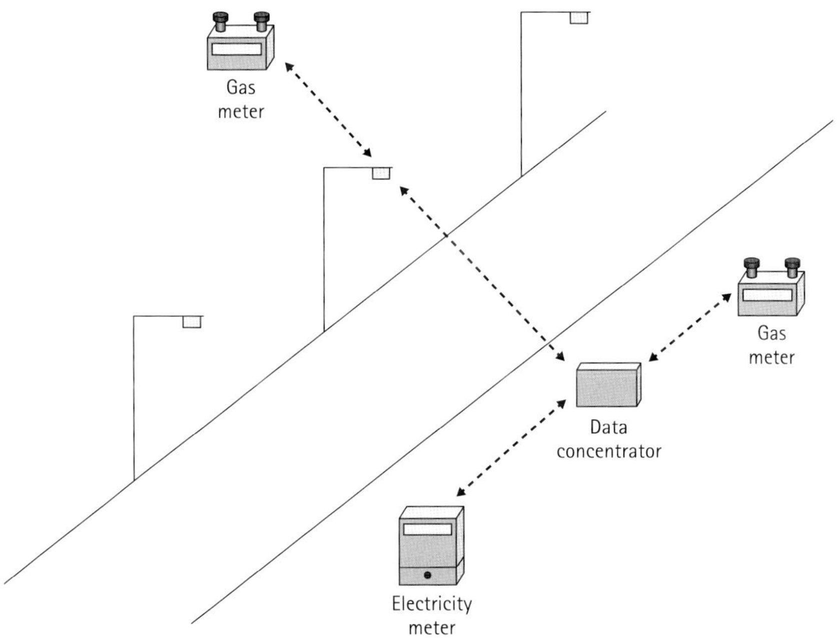


Fig. 5–1. A scenario using repeaters installed on streetlights

Thus, an obvious trend for smart metering systems is the creation of alliances between energy utilities and lighting service providers to enable a more cost-effective deployment of these platforms for sharing opportunities and benefits.

## Lighting automation

Nowadays, many lighting points on streets are switched using photoelectric receptors (fig. 5–2). These devices command lamps according to the level of available daylight. When artificial light is needed, the lamps are turned on; otherwise, they are off. This usually requires the installation of at least one device per streetlight post. An optional arrangement is the use of grouped switch panels. In this case, a central point is used to turn a group of devices on and off. An advantage of this is the reduction of clandestine connections on these circuits during the day, taking into account they are only supposed to be switched on at night. On the other hand, these points are also easy targets for vandalism because they can easily produce an area blackout. In addition to this, existing installations, based on photoelectric receptors, require a full grid reconfiguration and incur related costs.

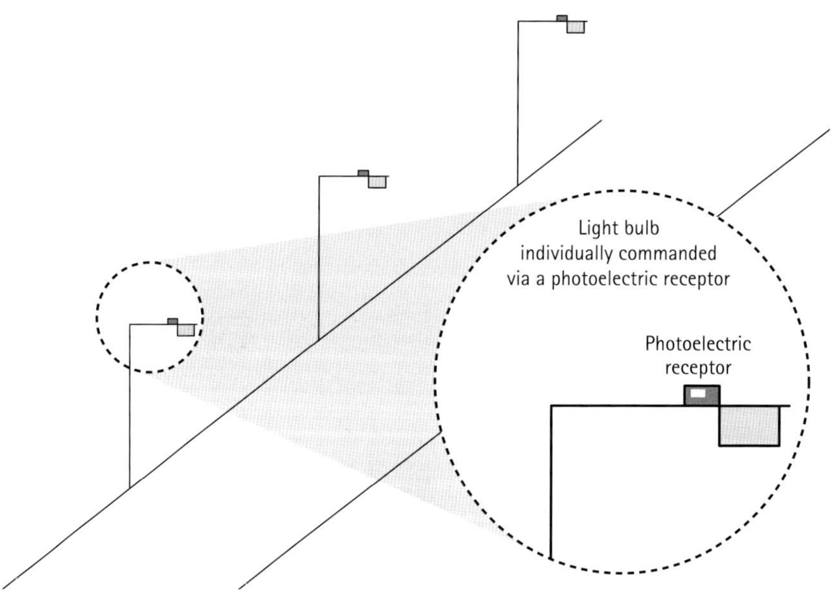


Fig. 5–2. An example of an existing streetlight configuration

A smart metering infrastructure could be used to deal with this challenge. For example, assuming a narrowband power line carrier (PLC) network, the data concentrator, in addition to its meters, could be used to command streetlights (fig. 5–3). A PLC device with an embedded switch could be connected to the lighting points and remotely commanded. Therefore, the data concentrator could send broadcast orders to command all the points simultaneously. These orders could be based on, among other sources, photoelectric sensors, timers, intelligent algorithms, or external information provided by weather forecasts. Middleware could schedule events for all the data concentrators according to the needs of the different areas. Note that the PLC medium was used only as an example. Other LAN communication technologies may also present the same potential for automation, such as low-power RF.

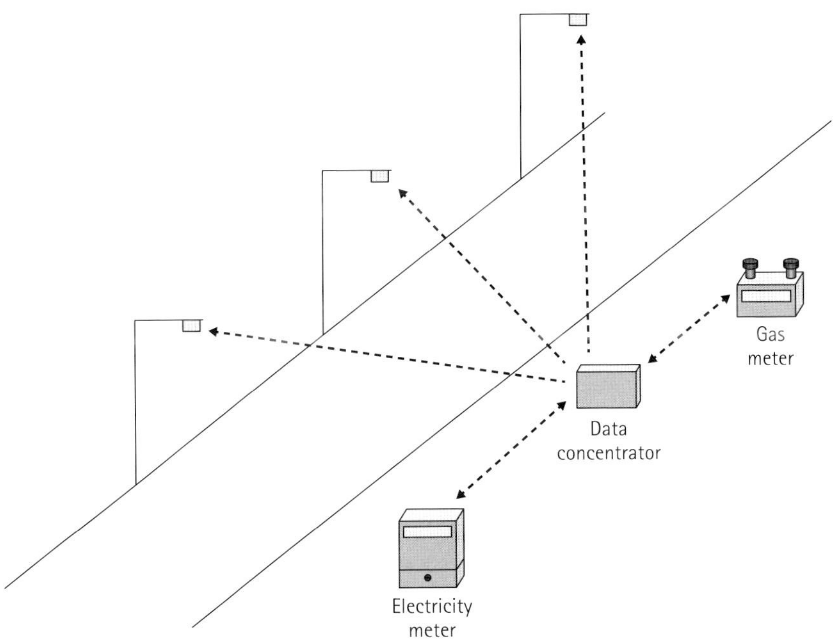


Fig. 5–3. Example of automated streetlights

Additionally, measurement circuits can be installed on these PLC devices. Apart from their switching abilities, this equipment could be used to measure streetlights individually, which is important for generating accurate billing data. As this equipment does not need to be individually read by customers, it is unlikely there will be any need for individual displays. Thus, the measurement circuit used for this purpose is generally very simple and inexpensive. Basically, it is a single board, size-reduced and without some common hardware used for electronic meters installed within customer homes such as liquid crystal display (LCD), connection terminals, covers, and so forth.

Technical and nontechnical losses associated with these circuits can also be reduced with the installation of meters. Beyond contributing to revenue protection, this can also improve safety for people and equipment. Individuals involved in illegal connections can be hurt or even killed, and these clandestine connections can damage both the grid and electrical devices due to bad electrical connections.

With existing grouped switch panels, a single electric smart meter or even a more simplified low-cost device could be installed in locations to command all the associated streetlights simultaneously. This would save costs for the number of devices needed to remotely manage public lighting, avoiding the need for reconfiguration of existing grid topology.

Assuming a more complex scenario, the deployment of smart metering systems may enable opportunities for a smart grid infrastructure for lighting circuits. For example, outage alarms could be generated, cross-referenced with other information, and analyzed. Thus, the service providers responsible for managing these circuits could benefit from automatic diagnostics about potential causes of faults and their locations. Many other lighting device faults could be easily detected for preventive and corrective maintenance purposes, such as flickering. Extra devices could be installed to automatically isolate faulty points and make prompt corrections.

Smart analysis may forecast the best time for replacements based on data such as lamp consumption over time, estimated life, and temperature variations. The same applications could be used to analyze load profiles to detect transitory faults such as bad contacts or lamps unexpectedly or briefly switched off (fig. 5–4). These functions are very useful for preventive maintenance.

Intelligent algorithms can optimize asset management. For example, specific lots of lighting equipment may present generalized anomalies that can result in extra costs. Asset databases and maintenance could be cross-referenced and, assuming anomalies in relation to life span prediction are detected within a lot, equipment could be isolated for treatment.

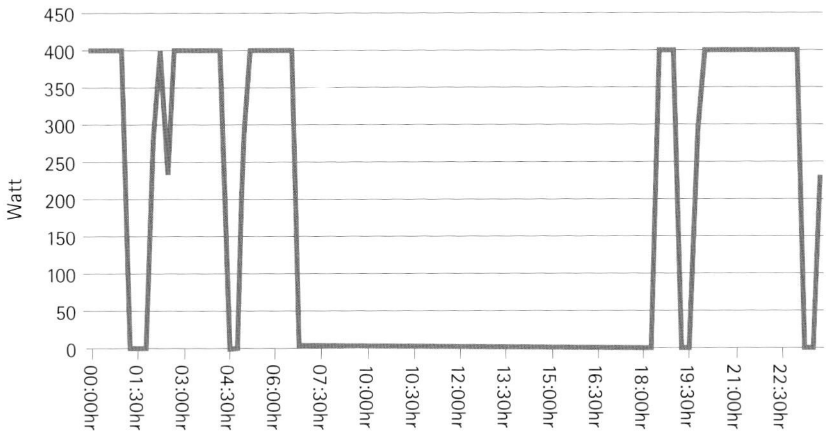


Fig. 5-4. Individual light-point profile showing temporary failures due to bad contacts

## Telecom opportunities

The use of streetlight infrastructure creates opportunities for RF mesh technology. For example, streetlight infrastructure could be used to improve the coverage of RF networks, acting as repeaters. Adding nodes related to streetlights extends the number of hops available to another device to communicate with a data concentrator, such as a smart meter. It also expands the overall coverage of these networks, for example, on streets. This may generate extra opportunities for energy utilities or lighting service providers. Assuming a massive rollout of smart streetlights, and taking into account that these lamps are everywhere, companies could be promptly informed the moment their vehicle arrives to attend a field event. On-board computers could be used to provide online text communication between technicians and the operation centers.

Amplifying this coverage may also have an impact on local maintenance. It extends the coverage of the smart metering LAN to the streets. Thus, in case of a wide area network (WAN) communication failure, handheld units could be used to communicate with smart meters located in customer homes.

With this LAN expansion, security and health care services could also be offered in public areas. For example, panic buttons could be installed at strategic points in streets, stores, or devices carried by technicians in the field. Traffic lights could benefit from this larger coverage and be monitored via the same data network. Electric and hybrid plug-in vehicles could also use these networks to communicate, as lighting points are likely to be installed in streets near parking locations (fig. 5–5). For example, drivers could be reminded of safety recalls and any other necessary maintenance by vehicle manufacturers or of recharging rate offers by utility companies.

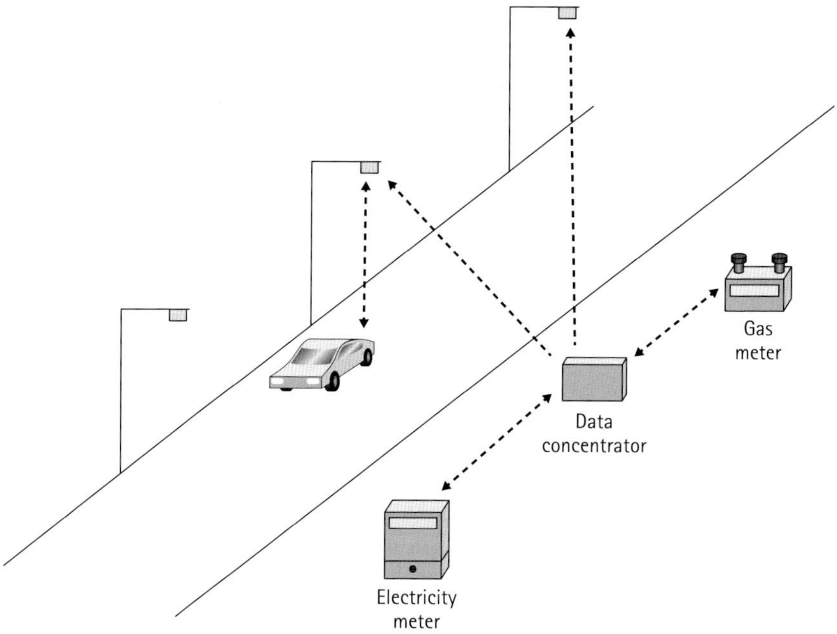


Fig. 5–5. A vehicle sharing the smart metering infrastructure with streetlight automations

## Definitively a trend

Based on all this, there is a clear trend to deploy smart metering systems with lighting automation. Synergies between energy utilities and these public service providers are numerous and may generate interesting alliances. As mentioned, the association of both platforms potentially increases opportunities and reduces risks through collaborative procedures and cost sharing. Utilities throughout the world are currently deploying pilots, and they tend to be upgraded quickly to become massive rollouts.

# Electric and Plug-in Hybrid Vehicles

Electric and plug-in hybrid vehicles seem to be the way forward for sustainable transport because of the potential reduction in $CO_2$ emissions, depending on the energy matrix of the country. The more renewable sources are used, the larger is the potential to reduce pollution levels. Electricity is appearing in the vehicle industry as an alternative to the fuels currently available in the market such as gas and diesel. Electric vehicles are fully independent alternatives, while hybrid plug-in vehicles combine traditional fuels with electricity.

Both types of vehicles can connect to AC sockets for recharging purposes. Beyond potential benefits to the environment, they offer opportunities to consumers in terms of transportation costs. Electricity tends to be cheaper than other available fuels in many countries. The savings may be even higher for customers if the recharging procedures are used during off-peak times. Energy is often cheaper and associated with lower sources of pollution at off-peak times.

Electric vehicles have been available for a long time, although problems persist such as the relatively short distances a vehicle like a car can be driven per recharge. Plug-ins represent the evolution of current nonrechargeable hybrid vehicles and could present solutions for this limitation. Thus, in addition to the electricity generated by the vehicle, the rechargeable version uses an extra, external AC source to supply power. These vehicles are also more economical than other conventional hybrid models in terms of fuel consumption and have extended range of use, as once the battery power is exhausted the internal combustion engine takes over.

Plug-in hybrid vehicles may create less $CO_2$ emissions than traditional vehicles and nonrechargeable hybrid vehicles. Tests on these vehicles began in 2007 in countries such as the United Kingdom, France, Belgium, Japan, and United States. Currently, these solutions seem to be more appropriate to urban rather than rural areas. These vehicles are currently able to run for 20 to 40 miles without recharging. They usually recharge in a few hours, and depending on the distance to be covered daily, more than one recharge may be needed during per day. Because of this, even if their adoption is becoming a reality, widespread use depends on energy utilities deploying more recharging stations. Although plug-in hybrids may be recharged via in-home sockets, public stations are needed as well. Thus, utilities should be ready to face this challenge, and smart metering platforms seem to be appropriate tools for this purpose.

Both types of vehicles may also provide very good opportunities for regenerating the automotive industry (fig. 5–6); this industry has been affected by the worldwide financial crises and the rise in oil prices. The new generations of plug-in vehicles could present a perfect opportunity to stimulate demand by replacing existing vehicles with this potentially green means of transport. However, there could be a serious collapse if this energy requirement is badly managed. For example, assuming 10% of vehicles in the market are recharging at the peak time, blackouts may occur. Contrary to expectations, $CO_2$ emissions may even increase depending on the sources of generation used to cover the deficit of electricity during this period of the day. Thus, demand-side management products that smart metering systems provide are essential to enable the secure entry of these new mobile consumers in the market.

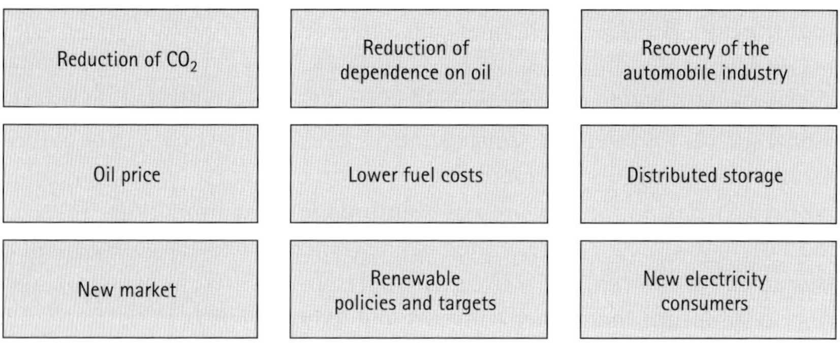


Fig. 5–6. Potential drivers for deployment of electric and plug-in hybrid vehicles

## Opportunities and drivers

Electric and plug-in hybrid vehicles offer opportunities for the environment, vehicle industry, and utilities. Some experts believe that millions of these types of vehicles will be deployed all over the world. Potentially, they could individually use hundreds of kilowatt-hours per month. Internationally, they could easily justify the building of new power stations to support consumption requirements.

Plug-in vehicles are potentially new sources of not only energy consumption but also distributed storage (fig. 5–7). Thus, they could actually be used to reduce peak consumption of electricity. They could store energy during off-peak periods and make it available to the grid when demand is highest. For example, early in the morning, customers could drive to work, where their vehicles could be recharged. Back at home, they could promptly plug their vehicles into electrical sockets, but delay the recharge (automatically) until later at night during off-peak periods. In the meantime, the remaining energy in the batteries could be converted to AC and used to supply the home during peak times. With this service, peak consumption on the grid could be significantly reduced. The extent of reduction depends on the number of vehicles deployed and available for this purpose within a utility area. In the future, with the evolution of new batteries for vehicles, more energy can be stored and later exported to the grid. Thus, these vehicles could become microgeneration sources on the grid.

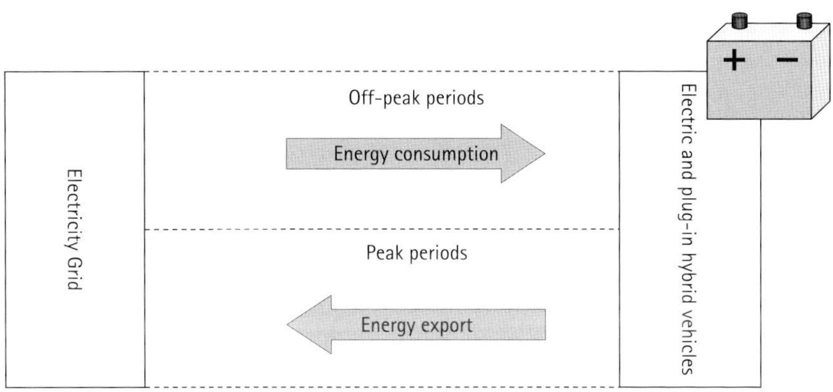


Fig. 5-7. Plug-in vehicles as electricity consumers and for distributed storage

In addition to benefiting from reductions in the price of fuel, customers have profited from many incentives provided by local governments. In Europe, for example, due to renewable targets driven by the Kyoto Protocol, some councils are providing free city parking in recharging areas and reducing taxes for these vehicles.

## Challenges

Battery technology has been a constant challenge for vehicle manufacturers. For example, the cost has been a problem, although it is getting close to resolution and has been decreasing every year. Despite this good news, battery size and storage capability are still a challenge. For example, currently plug-in hybrid vehicle models are only able to provide electricity to drive 20 to 40 per recharge. This seems to be compatible with urban area requirements but is unlikely to enable its free use for all applications. Despite this, technology is moving quickly, and more powerful and size-adapted batteries may soon be available on the market.

Another point is the availability of recharging stations. For the success of these implementations, recharging stations need to be widely available. As with other fuels, investors have already taken the lead to build charging stations in several countries. Despite this, energy utilities could play an important role here. In addition to the main stations, others could be available:

- Collective garages: While sockets located in private garages are generally supplied by the customer's meter, in collective garages such as in office and apartment buildings, they are generally provided via collective installations. Here, all the occupants, via the building rentals, would be sharing payment for the energy consumption required by these vehicles. Thus, it is necessary for utilities to provide ways for consumption related to recharges of vehicles to be attributed to the energy account of the vehicle owner, even if initially counted by another meter than the one owned by the customer.
- Public parking (fig. 5–8): In these areas, local authorities will offer technology compatible with the street environment and challenges such as vandalism and exposure to weather.

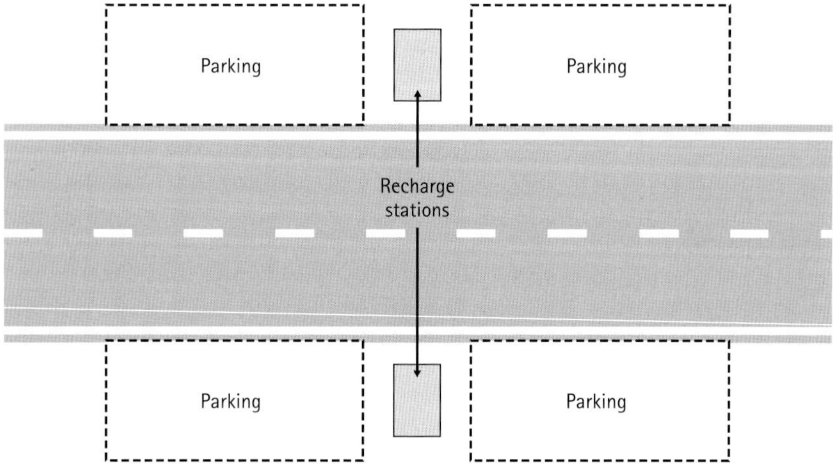


Fig. 5–8. Potential public parking configuration for electric vehicles using recharging stations

- Private parking: Special arrangements may be offered above the requirements of traditional private parking arrangements. For example, one multisocket station can be centrally located to supply four vehicles. Similar to the collective garages, this electricity could be attributed to the account of the consumer.
- In-home: Although electrical sockets are available in homes, plug-in hybrid vehicles require specific safety measures. For example, for recharging purposes, the ground cable must always be plugged in to avoid sparks and other problems related to static energy and isolation faults. Although many different mechanisms can be used for this purpose, these technologies are not available in the sockets currently installed in customer premises.

Safety procedures must also be developed. If poor arrangements are in place, the cables connecting the vehicles to recharging stations could create risks in public parking. Someone distracted could stumble on the cable and fall down. Thus, these arrangements must be deployed with efficient safety policies.

## Integration with smart metering systems

Recharging systems include more than just meters. They contain other components such as circuit breakers, recharge devices, IT platforms for energy management, and potentially HAN/WAN communication media. Analyzing these entities, the similarity with smart metering platforms is evident.

Smart metering systems offer a core role for these deployments. They can be integrated with vehicles, for example, to identify the users and redirect the energy consumed to their home accounts, independently of where the vehicle is recharged (fig. 5–9). Nowadays, vehicle keys and onboard computers store much information. They use industry standards and protocols to provide data exchange within the vehicles. An example of this is the controller area network (CAN) bus. CAN is a communication bus designed for industrial applications and deployed for vehicles since 1980. It is present in the majority of modern models. Experience demonstrates that CAN is easily integrated with external devices, including smart meters.

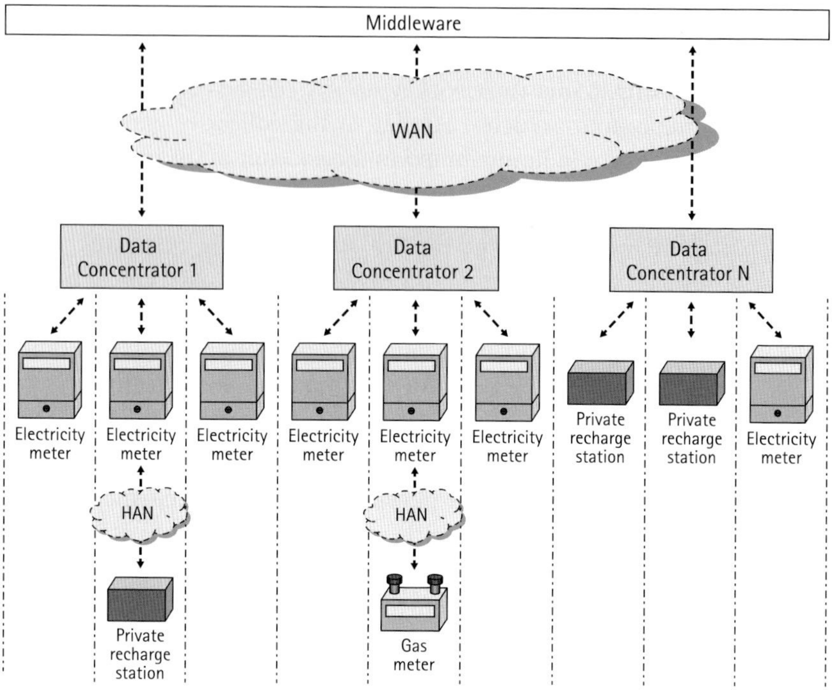


Fig. 5–9. Potential integration between smart metering gateways and plug-in hybrid vehicles

There are several options to enable integration in the field between smart metering systems and these sorts of vehicles. For example, smart meters could use HAN technology adopted by vehicle manufacturers. RF Bluetooth is currently been installed in many vehicles for integration with the external environment. Another option is to install smart bridge devices in vehicles to make them interoperable with the HAN managed by smart meters.

This integration enables continuous data exchange between vehicles and smart metering systems, at least during recharge events, for different purposes. Partnerships between vehicle manufacturers and energy utilities have already been established internationally to integrate both platforms. Integration enables recharges everywhere, under safety rules, and it amplifies the scope of opportunities for both energy utilities and vehicle manufacturers.

Logical integration could also be implemented in many ways (fig. 5–10). For example, the customer could create a vehicle account with their energy supplier. This account could then be linked with the customer's main account (related to their main residence) using a grouped rate structure, as already studied. For example, the customer's main residence could be associated with their vacation house, one electric scooter, and two plug-in hybrid cars. Thus, the customer would receive a single detailed electricity bill. Independently of where their vehicle was recharged, it would be identified by the smart metering system through a HAN via the smart meter used to supply the energy required (assuming an embedded in-home metering gateway). The energy measured locally could be stored in separated registers, logged, and then transferred from the meter to an account owned by the car owner. Thus, this special multimeter account would report consumption from different meters for billing purposes, and smart meters would be able to measure the energy charged to different customers.

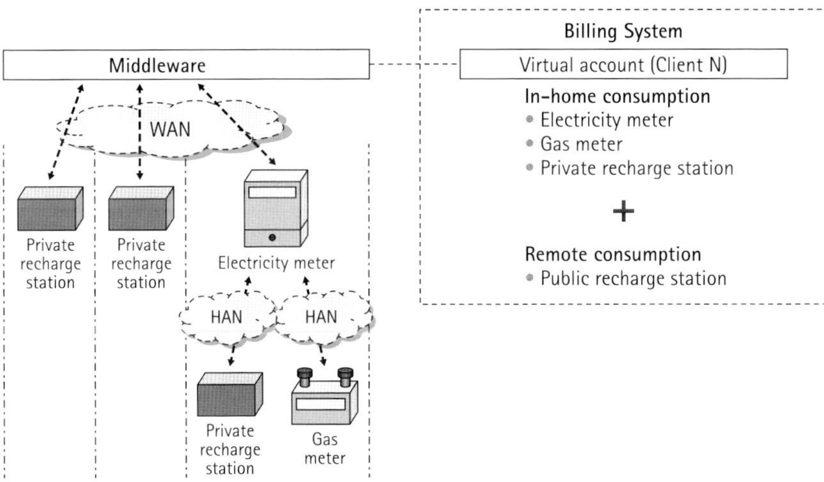


Fig. 5-10. Example of grouped rate associated with electric and plug-in hybrid vehicles

Unlike traditional consumers, these new consumers are mobile, so they will not be dependent on a single meter. This independence is necessary to enable the levels of comfort and flexibility required by these customers.

Beyond the grouped rate arrangement, another service seems very adaptable to these new consumers. Pay as you go (PAYG) is even more flexible with energy credit arrangements and the ability to buy energy anywhere. For example, the credit card information of the customer can be securely stored in a corporate database. Thus, customers may use the display interface in their vehicles to purchase energy credit locally. Similar to the previous arrangement, smart meters would update the customer's credit with consumption generated by multiple meters during recharging processes. This could enable an end-to-end integration starting in the vehicle, passing through the in-home metering gateway and middleware, and finishing in the energy utility applications. The opposite flow, from utility applications to the vehicle, would work as well. Scratch cards and Internet browsers could be interesting alternatives for credit purchase too. Green rates could be associated with this in order to help reduce $CO_2$ emissions.

The telecom link between vehicles and smart metering systems may also enable continuous safety checks while recharging (fig. 5–11). If any anomaly is detected (e.g., grounding interruption), the energy supply would immediately be stopped. Information monitored during this data exchange process could include connectivity to the ground, overcurrent and overvoltage detection, and other alarms.

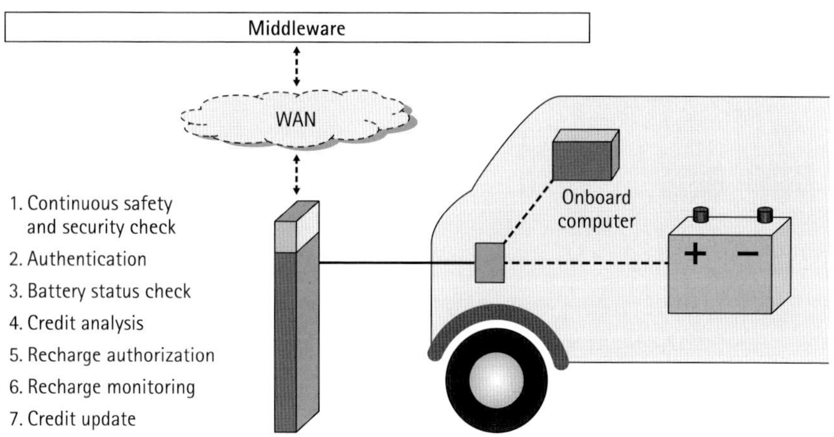


Fig. 5–11. A fully integrated recharging procedure for an electric vehicle

Data exchange can also direct recharging events to start and finish only after receiving orders directly from the vehicle. The vehicle's on-board computer could manage this procedure. The recharging process would only start after safety checks were conducted internally by the vehicle and externally by the utility recharging station; an allowance for recharging would be given by the energy supplier (e.g., enough credit in a PAYG account).

During recharging procedures, a lock mechanism could be used to block the plugs at both sides. Any plug removal would only be possible if the meter had already interrupted energy and there was no voltage on the recharging cable. Both cars and stations could also use sensors on their sockets to detect abnormal plug removals and quickly open the recharging circuit. This could be useful for reducing risks of vandalism and would avoid any kind of external electrical sparking produced during recharging procedures.

It would also be useful to know the status of the battery. When it is fully charged, a report could go to the meter that could promptly interrupt the flow of energy. The cable would remain plugged in but not connected to voltage to avoid energizing circuits when not in use.

A fully integrated scenario would allow vehicles to use all the resources of smart metering systems. For illustration, if at this stage a smart metering platform is integrated with the customer's television, the vehicle could also use it as a display. Thus, consumers could follow the recharging progress of their cars via their television. In the same way, they could receive messages and alarms such as marketing messages and alerts when their energy credit was low.

In addition, the vehicle manufacturer could use HAN interfaces or an interface within the vehicle to communicate with drivers. These interfaces could alert drivers to the following:

- Safety recalls
- Alarms (e.g., for abnormal plug removal)
- Preventive maintenance appointments or recommendations
- Marketing messages or messages from partners such as insurance, tire manufacturing, or maintenance companies
- Condition of vehicle components such as tires, oil filters, and wiper blades

Many other opportunities arise. Imagine the integration of these systems with a massive deployment in the streetlight industry; vehicles could be monitored and communicate with relevant companies almost everywhere. However, there are still challenges to be faced by energy utilities and vehicle manufacturers. Hard work is required for many market participants. Despite these challenges, these participants have enough expertise to overcome them.

Clearly, there are vehicle trends as well as strategic alliances between energy utilities and vehicle manufacturers. Energy utilities need to be ready for them and prepared to maximize potential opportunities.

# Smart Homes

Customer needs have been increasing with time. Every day new services are introduced to make our lives easier. Basic tasks can be carried out by machines, freeing up people to do more complex work. Most homes already have a level of intelligence and automation, such as:

- Air conditioning and heating systems are able to regulate temperatures according to environmental needs.
- Televisions are able to communicate with Internet modems and provide in-home entertainment.
- Fixed lines are shared for both telephone and Internet services.
- Remote controls are used to command appliances.
- Garage doors are automated.
- Alarm systems can detect and report intrusions.
- Appliances are programmable.

So, what is missing for these homes to become truly smart? Despite the level of automation, these components and systems continue to be isolated in the home. Interoperability seems to be the missing link. For example, alarm systems should be able to interact with garage door sensors. The alarm system should be able to detect their status and command the door to close when the owner had forgotten to close it and was not at home. Similarly, air conditioners and washing machines should be remotely operable from a display device such as a television or mobile phone. When achieved, it tends to generate massive deployments of smart homes internationally. Interoperability reduces the costs of automation deployments through its ability to share infrastructures from different in-home systems.

At this time interoperability may be seem to be a dream, but different drivers are moving the market in this direction, such as the evolution of technology, creation of standards, strategic alliances, less expensive components, availability, and mass deployment. For this reason, smart homes are definitively a trend for the medium term. This phenomenon presents opportunities to energy utilities for the potential integration of different in-home systems with smart metering platforms. Utilities should participate actively in the building process of in-home interoperability to ensure that their products are included. Different aspects must be developed, like interoperability limits, data security, and safety.

## Occupation management and safety

Assuming truly interoperable smart homes, sensors from different systems can be efficiently used to detect home occupation and safety to command in-home devices when necessary (fig. 5–12). For example, security platforms may use this information to automatically activate alarms when needed. Windows and doors could be then closed and locked, and premises-monitoring tasks activated. Water or gas leak detections and fire or smoke alarms could also be tuned to the maximum level of sensitivity. In case of a home invasion, local actions may be carried out such as lock reinforcement and alarm notifications. This platform could also promptly notify the police, security-monitoring companies, and consumers. In case of a fire, sprinklers could be activated, and firefighters, insurance companies, and clients would be promptly notified.

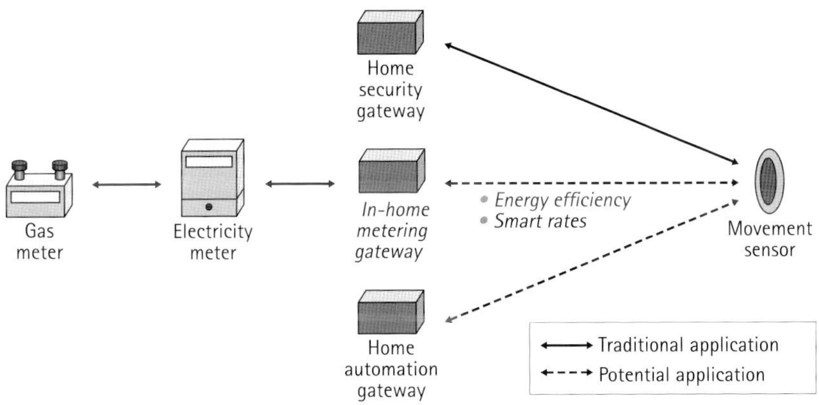


Fig. 5–12. Occupation management diagram—shared tasks

In more straightforward strategies, energy utilities could use their in-home metering gateway to provide the services described above. Otherwise, they could at least use this information to activate energy-efficient actions according to predefined profiles. If the security system detected that no one was home, lighting could be turned off. The same could apply to some appliances such as televisions and radios. Load-shifting services could be even more active during this period, too. In case a gas leak was detected or a stovetop burner had been left on, the gas supply could be switched off. Some actions could be performed after the customer was notified. Others, depending on the urgency, could be promptly executed followed by a report message. If needed, customers could confirm or reject orders from practically anywhere, with any source of WAN connectivity such as a mobile phone. In case of false notifications artificial intelligence could allow a system to learn from errors and avoid a repetition of the mistake in the future.

## Comfort and entertainment

Internet Protocol (IP) seems to be the trend in the medium- and long-term strategies of many authorities and telecom providers around the world. This is necessary internationally to provide increased comfort and to present opportunities for market differentiation through innovative offers to clients. It is also necessary to improve productivity. Globalization makes connectivity paramount; for example, it is now more common for people to work from home.

For these reasons, billions of dollars have been invested internationally in IP. Intensive research and development is being carried out to develop the best technical architecture design to achieve the best performance and maximize opportunities.

Beyond providing basic services, home entertainment providers are working with telecom and utility industries to offer combined services to provide more comfort. Sound and video streaming should to be available throughout the home and be controlled automatically according to customer needs (fig. 5–13). For example, wake up alarms can be integrated with news, weather forecasts, and traffic information streamed to TV, radio, or mobile devices. Plug-in vehicles could also be charged and available for use. When back home from work, both text and voice messages would be reproduced in audio form. Heating and cooling of the home could also be regulated according to a defined profile. Lighting scenarios could be defined according to different needs throughout the day and music set according to preferences. Doors and windows could be locked according to time and event settings, and electrical curtains could operate according to preference rules. Health care services could also be configured according to the residents' needs.

Taking into account the above scenarios, there are many opportunities for energy utilities. Interaction with these different systems would help energy utilities to understand consumer habits and apply energy-efficiency actions. Rates could be defined to match these behaviors according to the price of fuel and grid demand. Price signals could produce automatic tasks such as load shifting without strongly interfering with the levels of comfort required by customers. Potentially many new display interfaces could also be used. Different sources of in-home connectivity could enable utilities to piggyback metering and other relevant information using these infrastructures.

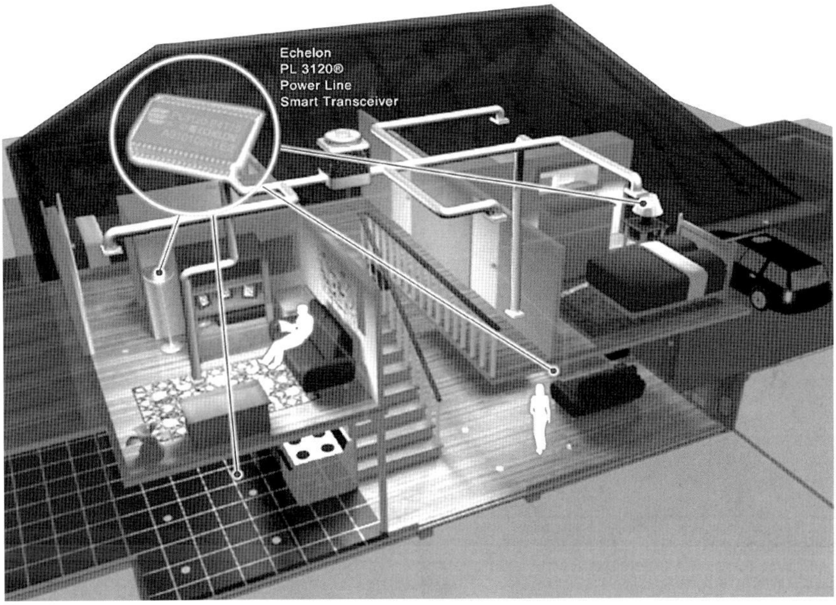


Fig. 5–13. Comfort and home entertainment diagram—lighting scenarios. (Illustration courtesy of Echelon Corporation.)

## Smart communicating appliances

New appliances with embedded HAN capability are currently available on the market (fig. 5–14). Examples are video games, televisions, refrigerators, and washing machines. Expansion for other types of appliances and mass deployment are just a question of time. Different market entities are defining an interactive environment where these different appliances could be integrated. In the integrated scenario, different applications could interact to improve overall performance in homes and reduce costs. For example, a television may act as a display device used to command washing machines and ovens. Appliances could be switched on and off by scheduling events or on-demand orders. These tasks could be locally or remotely managed via secure web-based applications.

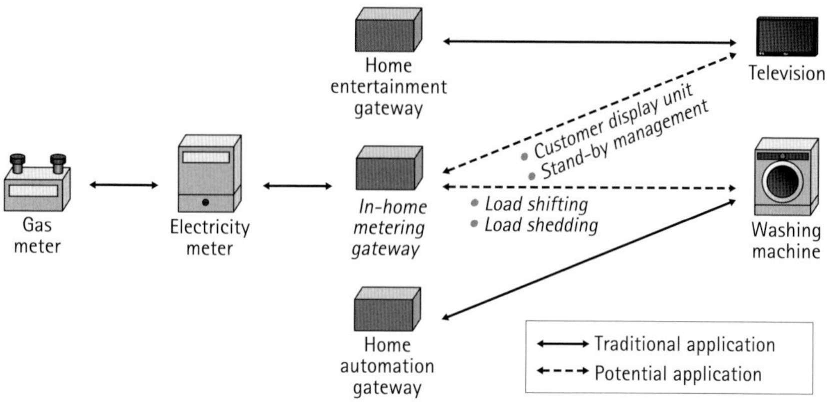


Fig. 5–14. Smart appliances diagram—shared tasks

Energy utilities could make good use of these structures. Televisions could be turned off when the room is vacant for a defined amount of time. Washing machines could operate when the price of energy and carbon emissions are lowest (fig. 5–15). Refrigerators could be switched on and off according to load-shifting arrangements. In summary, demand-side management would become more efficient and cost-effective. Utility gateways could interact directly with controllers and switches embedded in these appliances. Apart from cost improvements, energy-efficiency actions would take place according to authorization given by the appliance. For example, perhaps a device should not be switched off at a specific time because an important internal task is being executed. An external switch would be unlikely to know this.

## Smart metering integration

Analyzing the trends and the utility opportunities, smart metering systems are likely to be integrated with smart home platforms. Energy utilities should be prepared to consider this trend in their potential strategy plans and business cases because of its associated benefits.

Home environments need a device manager. It must supervise and control all the in-home applications and provide interoperability among applications based on different standards and media. It should also link with the WAN for flexibility within the home. As you are now aware, smart metering system platforms could potentially provide a device with full controls, depending on how it is specified. Thus, in-home metering gateways can play an important role in this integration process. Due to the gateway's strategic position within the home, partnerships with home controller providers seem to be the next logical step for many energy utilities, subject to legal requirements (fig. 5–16).

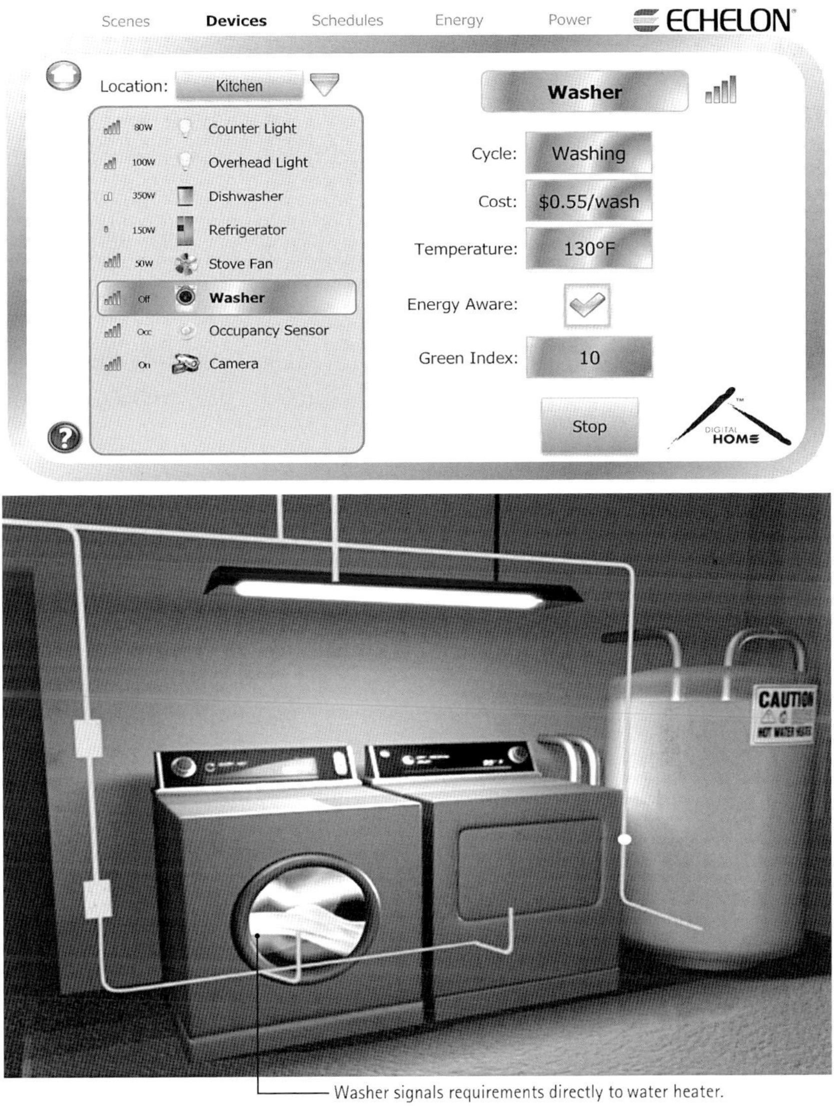


Fig. 5–15. Smart communicating appliances diagram—shared tasks. (Image courtesy of Echelon Corporation.)

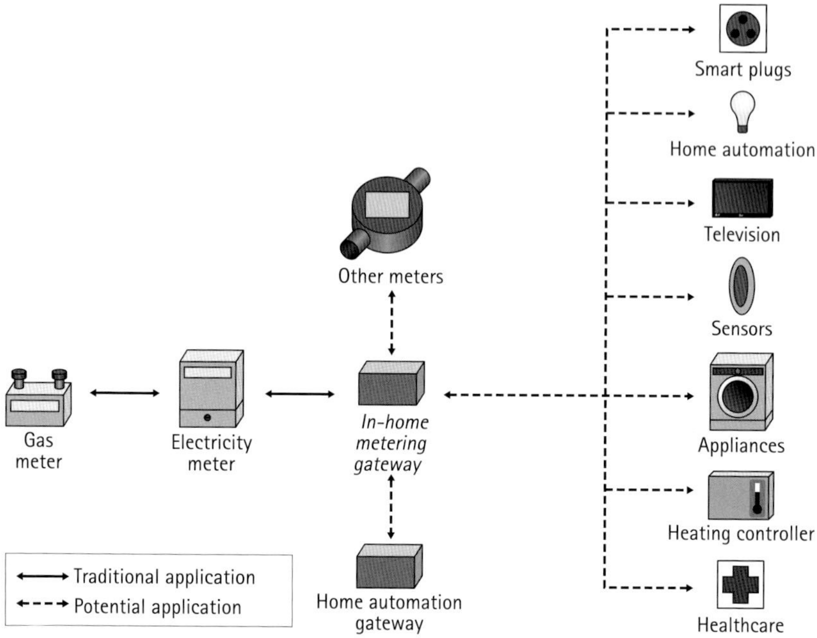


Fig. 5-16. Potential integration of smart metering systems with in-home applications

This integration offers the possibility of sharing information and infrastructure with components in the home. It makes life easier for utilities because of the possibility of interventions directly at the load source. Sharing these components also means using consumer interfaces already consolidated with other applications.

The smart home is a concept that consumers find fascinating, but ideas are sometimes not put in practice because of the costs involved. This barrier could be mitigated by the savings from sharing infrastructure and the opportunities of deploying both platforms together. The smart home would consequently reduce the risks of implementing smart metering projects because of the added functionality value for customers. Home automation has demonstrated that customers are prepared to pay for it if the cost is fair and the benefits meet expectations. The expense could at least be partially financed by customer savings from reduction of consumption and better management the customer's budget. Thus, the smart home seems to be an irresistible model, and for this reason, energy utilities around the world have been promoting it in competitive markets.

# Smart Grids

Smart metering systems can be stepping stones to smart grid projects. They offer the opportunity to share infrastructure and reduce the cost of smart grid projects. For example, electricity meters may act as sensors on the grid because of their capability for advanced quality measurements and online feedback.

## The challenge

The grid is one of the main assets of electricity utilities, but managing it without automation is a challenge. This is due to many factors:

- Diversity of environments, such as underground and aboveground feeders
- Variations of population density
- Size of service areas
- Areas that are difficult to access
- Number of consumers
- Weather conditions
- Distance between operational centers and cities
- Susceptibility to environmental interference, such as contact between feeders and tree limbs or roots, and traffic accidents causing damage to power poles and the grid
- The need for corrective and preventive maintenance in the field
- Dependence on customer feedback to identify specific types of faults on the grid

Even with this evidence, few networks in the world today are fully or even partially automated. Cost is generally the main obstacle for these deployments. Among others, communication infrastructure may be very expensive given the locations of the targeted devices around the utility area. Often, these points are not centralized, which increases the costs to deploy telecom infrastructures. Additionally, telecom coverage may not be adequate to provide connectivity to all the strategic devices, or might be completely unavailable in some areas. Because of this, despite the benefits, it is unlikely that these projects will be economically justifiable by themselves. However, when associated with smart metering projects, they can provide good payback in most cases, as a result of reduction of costs related to sharing common infrastructures. This is the reason why more and more smart metering projects are designed with the deployment of smart grid systems. For example, LAN-based architectures may be chosen to share this local telecom network between both projects. However, interoperability between both platforms needs to be established.

Although interoperability is a challenge, international research and development efforts have created new architectures and standards for smart grids. For example, in the United States, IntelliGrid is an Electrical Power Research Institute (EPRI) initiative to create technical foundations to guide utilities on interoperable smart grid projects. EPRI is a nonprofit organization that conducts research and development on technology, operations, and environment for the global electrical power sector. The GridWise Alliance is a consortium with similar objectives, formed by public and private companies. It provides a forum for technologies, methodologies, and many other factors. In the U.S., efforts are underway to normalize the different elements that can compose smart grid projects. The National Institute of Standards and Technology (NIST) has established a framework to ensure interoperability between the different smart grid technologies.

Many other initiatives are being managed by utilities and partners to make this technology a mass reality.

## Solutions, opportunities, and drivers

Basically, smart grid systems may help utilities to:

- Anticipate disruptions. Smart grid devices associated with smart metering are able to provide online information about network status and enable accurate, short-term usage forecasts. Disruptions could be avoided with demand response action or power limitations. In extreme cases the impact could even be reduced via load curtailment.

- Reduce operational costs of the grid. As discussed, due to its topology and other characteristics, grid management is very expensive for electricity utilities. Field visits by technicians make these costs even higher. These costs tend to be reduced with the new smart platforms thanks to the level of automation these systems can provide.

- Enable renewable and distributed generation. Smart management allows the energy matrix to be cleaner due to reduction of peak consumption and mass deployment of renewable micro-generation sources and less dependence on high-pollution power plants. Distributed storage points could also be installed around the network to optimize daily consumption curves.
- Improve dispatch. Online monitoring and self-reconfiguration are likely to help utilities optimize their dispatch tasks.
- Prepare the grid for the next century. Asset modernization and automation are necessary to enable utilities to handle challenges, including the trends of increasing demand. Otherwise, these demand levels can reach a point where the amount of energy required is more than what can be generated. This could cause blackouts for whole cities and states. Some utilities are already suffering from this problem.
- Improve quality of supply. Voltage regulation, reactive energy, harmonics management, and phase balancing are just some examples of the opportunities these projects may offer to improve quality of supply.
- Improve reliability and reduce outages. These platforms enable grids to offer better performance for utility consumers. Damage to the grid caused by downed trees and tree limbs is one of the major contributors to outage events in some countries. These platforms enable maintenance teams to strategically plan tree pruning.

With the integration of these platforms and smart metering systems, meters can act as sensors and switches for the grid. However, other devices also need to be installed, for example, defect-detector sensors and automated high-power switches. These devices can be used to automatically identify and isolate faults on the electricity network.

In an unautomated environment, where a fault creates a network protection operation, and consequently a local outage event, technicians must go into the field. They need to run through the entire network to identify the problem, open switches at different locations to isolate the fault, and reestablish unaffected points. This procedure may take hours and affect consumers for a longer period than it should. In an automated environment, the problem may be identified and isolated in seconds, with automatic network reconfiguration (fig. 5–17). Therefore, technicians could go straight to the points where corrections are needed.

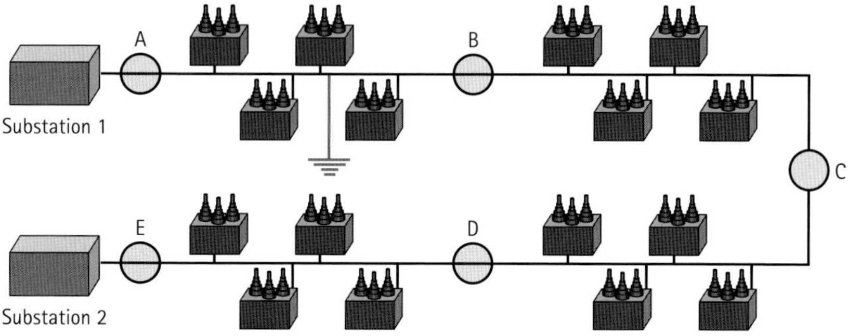


Assuming a fault between switch A and B, the circuit could be automatically reconfigured in order to isolate the fault and re-establish the energy for the non-affected points.

For example, switch A and B could be opened and switch C closed. In this case, the last four transformers, originally supplied by the substation 1, would be temporarily supplied via substation 2.

**Fig. 5–17.** Example of fault isolation

Preventive maintenance and other actions could also be based on data generated by smart meters. This provides the opportunity to send feedback to the network operators for outage detection and energy-quality monitoring. The analysis of these data could be even more beneficial if associated with other strategic devices, for example, from fiscal meters installed at power transformers. This could help avoid future costs of faults originated by poor management of load on these transformers. Instead, detection by smart devices of abnormal phase balancing, peak overloads, and general overcharges makes it possible to take preventive action before a fault or outage occurs. Many other optimizations could be facilitated by integration between the smart grid and smart metering platforms, depending on the specific needs of energy utilities.

## Smart metering systems: The port of entry for smart grid

It is unlikely that smart grid systems will be deployed in isolation. They are likely to be deployed by piggybacking on a smart metering infrastructure to enable devices to communicate and interact. But there will be challenges to face. For example, standards need to be created to facilitate interoperability.

Given available synergies, both platforms are likely to be integrated to amplify their scope. For example, demand-side management systems could be associated with demand-response programs based on grid needs. The opposite is also true; energy balancing from meters measuring substation feeders and power transformers may also help metering initiatives, such as revenue protection programs. Integration is beneficial for both sides.

The integration of both platforms is already becoming a reality for some utilities around the world. Their integration and massive deployments are becoming a trend in the medium and long term.

# Conclusion

Smart metering platforms are no longer a trend, but a reality across the world. Although there are still some challenges and new methodologies ahead, most have been mastered by energy companies.

These technologies will be, sooner or later, massively deployed worldwide. It is just a question of time, as these systems offer many benefits to customers, energy utilities, environment, local authorities, and other market participants.

The integration capability of these platforms amplifies the potential scope of business for energy utilities and transforms energy into more than just a commodity; it is also a source of profits from new products and services offered to customers. These profits could also come from optimization of existing operational processes and the consequent reduction of expenses and overhead. These are only some examples; the benefits these platforms can offer are unlimited.

Many drivers and opportunities come with these initiatives. These aspects should be studied in detail for each market. Business models and evaluation exercises should take each available opportunity into account when investigating the potential benefits for all market participants.

## Smart Metering Progress: The Big Picture

Taking into account an international market with hundreds of millions of customers, it is difficult to imagine efficient management of all these consumers without smart metering platforms. Competition in energy markets is also becoming more aggressive, with potentially millions of customers changing energy suppliers every month internationally.

These are some of the reasons why many energy utilities around the world are massively deploying smart metering systems. Others are preparing their requirements and investigating the potential for generalizing these platforms. A few have not yet woken up to this reality, but sooner or later they will realize that smart metering is inevitable. Hundreds of millions of smart meters are expected to be introduced to the residential market within the next 10 years across all continents.

International benchmarking demonstrates that the way these rollouts will be organized depends on different aspects, mainly related to the local environment and strategies of the different market participants involved. In order to determine the best strategy to be implemented, some questions should be answered:

- What do customers think about smart metering? Are they currently interested in these platforms?
- How can customer behavior be influenced through this technology?
- How do customers benefit from this technology? Do they understand the benefits?
- What obstacles need to be overcome for these platforms to be delivered?
- Is there a commercially viable business case for all the market participants and segments?
- What are the opportunities these platforms offer to a specific region or range of customers?
- How difficult is it to integrate these platforms with others in order to expand benefits and opportunities?
- Will governments and other local authorities incentivize and encourage this new technology?
- How can a smart metering rollout be implemented in a fully competitive market and be used by all energy suppliers without extra operational costs, such as meter changes every time a customer changes supplier?
- Can all the different energy suppliers of a competitive market work together to provide a common framework? How will the regulatory agency or government view this with respect to competition?

This book provides information that may help you deal with these questions and many others about the deployment strategy of smart metering systems. All these aspects must be carefully considered by project teams, along with their requirements, strategies, and background. The technical teams of energy utilities and other specialists in the market are certainly able to deal with these challenges as well as the potential opportunities.

## Defining the Solution

Defining a technical architecture is fundamental to the success of smart metering systems. Apart from the need to safely and securely specify hardware and software, a technical architecture is necessary to enable integration with other platforms and interoperability between all the devices within its scope.

Experience from past initiatives should be the basis for innovation. It is important to understand the evolution of metering systems over time and their purpose. This experience will help you develop robust platforms. Benchmarking exercises are fundamental for this learning exercise, but there is no single solution. In general, some customization is needed for any implementation.

It is also important that a smart metering system will be comprised of a complete, interoperable, and open platform in terms of meters, communications, systems, and protocols that can be used by the different market participants and ensure future-proofing for customers, market needs, and trends.

For this reason, interoperability features must be well established, allowing relationships among the different devices in the field, such as gas meters, electricity meters, customer displays, other devices, and their IT management platforms. This concept is not restricted to technology. For example, customers should also actively interact with the technology. Their participation is crucial to achieving the objectives of these projects. Their feedback should be taken into account to facilitate the rollout of efficient automatic products and services by energy utilities and to help customers realize their savings goals.

Different solutions are available in the market that claim to be smart metering systems. They offer different functionalities and use different technological options such as communication media, protocols, and measurement methods. The chosen technology must be fully compatible with the utility strategy, customer needs, local laws, and regulatory policies.

That is why, although there are a large number of offers available in the market, it is hard to find a product that is completely compatible with specific market needs and requirements.

Different standards have also been implemented across the world, based on local needs, laws, and regulations. Various participants have tried to integrate these standards and customize them into specific market requirements. The success of smart metering deployments is directly dependent on these specifications.

Data security, data management, and other IT parameters are also necessary for these implementations. Policies must be reviewed in regard to these parameters. Additionally, safety and legal criteria are paramount.

For all these reasons and many others discussed in this book, energy utilities must take special care when building these solutions. The cycle of transforming requirements into technical solutions is complex but achievable if the correct methods are used.

# What's Next?

Smart metering teams should closely watch market trends. At the least, integration with other platforms can amplify the range of benefits to be achieved. There are also other projects that could only be justified if implemented in association with smart metering systems, such as smart grid platforms. Therefore, a global view is necessary to realize all the benefits and opportunities for the different market participants.

Although this book predicts some trends, what is next depends on you and other parties directly or indirectly involved in projects. Continuous evaluation is necessary as trends evolve.

Finally, I hope this book offers you some elements to help you deal with smart metering systems in a better way. I hope you make good use of this book, and I hope to meet you personally, someday, somewhere—maybe in one of the many forums around the world, to discuss this theme. Together we can help to build the future. Thus, it would not be appropriate to say goodbye; instead I say: 'bye for now, hope to see you soon.

# Glossary

## List of Acronyms

| | |
|---|---|
| **A** | ampere |
| **AC** | alternating current |
| **AMI** | advanced meter infrastructure |
| **AMM** | automatic meter management |
| **AMR** | automatic meter reading |
| **ANSI** | American National Standards Institute |
| **ATM** | automatic teller machine |
| **B2B** | business to business |
| **B2C** | business to client or consumer |
| **bps** | bits per second |
| **CAN** | controller area network |
| **CDMA** | code division multiple access |
| **CENELEC** | European Committee for Electrotechnical Standardization |
| **CRM** | customer relationship management |
| **CSD** | circuit switched data |
| **DC** | direct current |
| **EEPW** | EDF Energy Project Way |
| **EDF** | Électricité de France |
| **EPRI** | Electrical Power Research Institute |
| **ERA** | Energy Retail Association |
| **ESCO** | energy service company |
| **GIS** | Geographic Information System |

| | |
|---|---|
| **GPRS** | general packet radio system |
| **GPS** | Global Positioning System |
| **GSM** | Global System for Mobile |
| | |
| **HAN** | home area network |
| | |
| **IEC** | International Electrotechnical Commission |
| **IEEE** | Institute of Electrical and Electronics Engineers |
| **ISO** | International Organization for Standardization |
| **IT** | information technology |
| | |
| **LAN** | local area network |
| **LCD** | liquid crystal display |
| | |
| **MBUS** | Meter-Bus |
| **MMS** | multimedia message service |
| | |
| **NAT** | network address translation |
| **NAN** | neighborhood area network |
| | |
| **OEM** | original equipment manufacturer |
| **OFDM** | orthogonal frequency-division multiplex |
| | |
| **PAYG** | pay as you go |
| **PCB** | printed circuit board |
| **PET** | polyethylene terephthalate |
| **PLC** | power line carrier |
| **PSTN** | public switched telephone network |
| | |
| **RF** | radio frequency |
| **RFI** | request for information |
| | |
| **SCADA** | supervisory control and data acquisition |
| **SIM** | subscriber identity module |
| **SLA** | service level agreement |
| **SMS** | short message service |
| **SRSM** | Supplier Requirements for Smart Metering |
| **STS** | standard transfer specification |
| **SWOT** | strengths, weaknesses, opportunities, threats |

**TCP/IP** Transmission Control Protocol/Internet Protocol
**TDMA** time division multiple access

**UML** Unified Modeling Language
**USB** universal serial bus

**V** volt
**VA** volt-ampere
**var** volt-ampere reactive
**VHF** very high frequency
**VoIP** voice over Internet Protocol
**VPN** virtual private network

**W** watt
**WAN** wide area network
**WAP** wireless application protocol
**WiMax** worldwide interoperability for microwave access

# Definitions

**competitive.** Possessing the necessary capabilities to compete.

**fossil fuel.** Fuel (as coal, oil, or natural gas) formed in the earth from plant or animal remains.

**high voltage.** Called high-voltage B (HTB in France), high voltage is defined as the voltage range used for the transportation of electricity. It is between 50,000 and 400,000 volts. More specifically, the range between 225 kV and 400 kV is called very high voltage (THT in France). These values and nomenclature vary depending on country standards.

**low voltage.** Between 230 and 400 volts.

**maintenance.** All operations required for support, verification, and repair.

**medium voltage.** Called high-voltage A (HTA in France), medium average voltage is defined as the voltage range used for the distribution of electricity. It is between 1,000 volts (1 kV) and 50,000 volts (50 kV). These values and nomenclature vary by country, depending on national standards.

**nonrenewable energy.** Energy generated from a resource that is fixed or limited (e.g., oil, coal, gas, and uranium).

**photovoltaic.** Technology that enables the production of electrical energy from solar radiation.

**power.** Energy produced per unit of time. Power is measured in watts.

**pressure.** Force exerted by a substance (liquid, solid, gas) on another in direct contact with it.

**renewable energy.** Energy derived from resources that are renewed continuously in nature (e.g., water, wind, sun).

**voltage.** Potential difference between the positive terminal and negative terminal of an electrical device. Voltage is measured in volts with a voltmeter.

# Bibliography

## Chapter 1

Baker, R. C. (2000). *Flow Measurement Handbook: Industrial Designs, Operating Principles, Performance, and Applications.* Cambridge, England: Cambridge University Press.

Creder, H. (1994). *Instalações Elétricas.* Rio de Janeiro, Brazil: Livros Técnicos e Científicos.

De Medeiros Filho, S. (1997). *Medição de Energia Elétrica.* Rio de Janeiro, Brazil: Livros Técnicos e Cientificos.

*Deteccao de Fraudes em Medição Indireta.* Rio de Janeiro, Brazil: Centro de treinamento Operacional Engenheiro Edmilton Zaroni.

Di Jora Guedes, C. *A história dos Medidores de Energia Elétrica.* Rio de Janeiro, Brazil: Light Serviços de Eletricidade.

Edison Electric Institute. (2002). *Handbook for Electricity Metering* (10th ed.). EEI Publication.

*Inspecao e Aferição em BT sem Transformador de Corrente.* Rio de Janeiro, Brazil: Centro de treinamento Operacional Engenheiro Edmilton Zaroni.

Miller, R. W. (1996). *Flow Measurement Engineering Handbook.* New York: McGraw-Hill.

Wiebe, M. (1999). *A Guide to Utility Automation: AMR, SCADA and IT Systems for Electric Power*. Tulsa, OK: PennWell.

# Chapter 2

Akass, C. (2007). "Smart Meters Could Bring Free Home Hubs." *Computer Active*. Accessed 25 June 2008 at: http://www.pcw.co.uk/personal-computer-world/news/2196563/free-hubs-smart-meters-3359109

Carvalho, L. A. V. (2005). *Data mining: A Mineração de Dados no Marketing, Medicina, Economia, Engenharia e Administração.* Rio de Janeiro, Brazil: Ciência Moderna.

Edison Electric Institute. (2002). *Handbook for Electricity Metering* (10th ed.). EEI Publication.

Forouzan, B. A. (2006). *Comunicação de Dados e Redes de Computadores.* Porto Alegre, Brazil: Bookman.

Miller, R. W. (1996). *Flow Measurement Engineering Handbook*. New York: McGraw-Hill.

Tanenbaum, A. S. (1997). *Redes de Computadores.* Rio de Janeiro, Brazil: Campus.

Tanenbaum, A. S. and M. V. Steen. (2007). *Sistemas Distribuídos.* São Paulo, Brazil: Prentice Hall.

Shaw, A. C. (2003). *Sistemas e Software de Tempo Real.* Porto Alegre, Brazil: Bookman.

Wiebe, M. (1999). *A Guide to Utility Automation: AMR, SCADA and IT Systems for Electric Power*. Tulsa, OK: PennWell.

Series of articles presented by the author and by third parties at different international technical congresses organized by Spintelligent and/or Synergy-events available at www.metering.com listed below:

- Metering Latin America, 2005, 2006, 2007, 2008, 2009, and 2010
- South Africa Pre-payment Week, 2006
- Metering, Billing and CRM Asia-Pacific, 2006
- Metering Europe, 2008, 2009
- Grid Week 2010, United States of America
- IRIS 2010, Renewable Energy Storage Conference, Germany
- Smart Grid Brazil Forum 2010

- Smart Electricity World Latin America 2011
- Cidades Verdes 2011, Brazil

# Chapter 4

Baker, R. C. (2000). *Flow Measurement Handbook: Industrial Designs, Operating Principles, Performance, and Applications*. Cambridge, England: Cambridge University Press.

Edison Electric Institute. (2002). *Handbook for Electricity Metering* (10th ed.). EEI Publication.

Miller, R. W. (1996). *Flow Measurement Engineering Handbook*. New York: McGraw-Hill.

Wiebe, M. (1999). *A Guide to Utility Automation: AMR, SCADA and IT Systems for Electric Power*. Tulsa, OK: PennWell.

Series of articles presented by the author and by third parties at different international technical congresses organized by Spintelligent and/or Synergy-events available at www.metering.com listed below:

- Metering Latin America, 2005, 2006, 2007, 2008, 2009, and 2010
- South Africa Pre-payment Week, 2006
- Metering, Billing, and CRM Asia-Pacific, 2006
- Metering Europe, 2008, 2009
- Grid Week 2010, United States of America
- IRIS 2010, Renewable Energy Storage Conference, Germany
- Smart Grid Brazil Forum 2010
- Smart Electricity World Latin America 2011
- Cidades Verdes 2011, Brazil

# Index

## A

## B



## C

## D

## E

## F



## G

## H



## I

# M

## N



## O

# S

## T



## U

## V



## W



## Y



## Z

# About the Author

**Fabio de Oliveira Toledo** is the president of SGI Consulting. Up to January 2013, he worked for Light, a former subsidiary of Electricité de France (EDF) in Rio de Janeiro, Brazil, as chief technology officer (CTO) and executive coordinator of the smart grid program. From 2005 to 2009, he was expatriated within the EDF Group, at EDF R&D in France and in the United Kingdom, where he was responsible for managing smart grid and smart metering R&D projects. Prior to his expatriation, his last position was director of the Metering and Revenue Protection Department f or large accounts at Light.

He is author of books about smart metering and smart grids (published internationally), specialization professor, and has more than 15 years of managerial and technical experience in energy markets. He has a solid technical education background: an MBA update in international and strategic management from the University of Grenoble in France and the London School of Business and Finance (LSBF) in the United Kingdom, an MBA in business management in energy markets from Instituto Brasileiro de Mercado de Capitais (Ibmec-RJ) in Brazil, another MBA in IT business management from Centro Federal de Educação Tecnológica Celso Suckow da Fonseca (RJ-CEFET) in Brazil, and a postgraduat e Latu Sensu in Internet, interface, and multimedia at Universidade Federal Fluminense (UFF) in Brazil.

He is multilingual (Portuguese, French, Spanish, and English) and enjoys broad international recognition in the metering arena. He has written articles in specialized publications, participated in technical conferences (chairman, panelist, speaker), been interviewed in the media (newspapers, radio, television) and participated in several work assignments abroad (America, Europe, Asia, and Africa).

Toledo is also recognized internationally for his work on metering standardization. He represented the companies for which he worked in several international working groups of standards and norms: International Electrotechnical Commission (IEC), European Committee for Electrotechnical Standardization (CENELEC), The Application Home Initiative (TAHI), Associação Brasileira de Normas Técnicas (ABNT). In this framework, he was a member of the workgroups TC 13 (IEC-CENELEC), Interoperability Framework Initiative (IFI), TAHI, and Suppliers' Requirements for Smart Metering (SRSM) Project Energy Retail Association (ERA). He is also the author of several corporate specifications.